SLAYING THE SKY DRAGON

DRAGON

Death of the Greenhouse Gas Theory

The Settled Climate Science—Revisited

www.slayingtheskydragon.com

This book is dedicated to Ernst-Georg Beck and all the like-minded scientists around the world.

The authors' personal acknowledgements can be found at the end of the book.

ISBN 978 0 9827734 1 3

STAIRWAY PRESS

1500A East College Way #554
Mount Vernon, WA 98273
www.StairwayPress.com

CHAPTER INDEX

Tim Ball

Alan Siddons

Tim Ball

Martin Hertzberg

Hans Schreuder

Joe Olson

Claes Johnson

Charles Anderson

John O'Sullivan

Disclaimer

The authors of this book received no funding from government agencies or private corporations and the opinions expressed are entirely their own.

Chapter 1

Analysis of Climate Alarmism—Part 1

by Tim Ball

Introduction

THE MOST FUNDAMENTAL assumption in the theory that human CO_2 is causing global warming and climate change is that an increase in CO_2 will cause an increase in temperature. The problem is that in every record of any duration for any period in the history of the earth exactly the opposite relationship occurs: temperature increase precedes CO_2 increase. Despite that a massive deception was developed and continues.

How did the massive deception of human induced global warming bypass the normally rigorous scientific methods? Why does it continue to survive? Who orchestrated the science and the politics? What was the motive?

Two major factors explain how the Anthropogenic Global Warming (AGW) people got away with massive deception. First was exploitation of fear. The end of the world is coming, there's only a few years left became the mantra of everyone from UN Secretary General Ban Ki-moon to Prince Charles. Second was exploitation of people's lack of knowledge or understanding of science. This is more easily exploited because of the distribution of people that understand science and those who have no idea and are often proud of the fact. After twenty-five years of teaching a science credit course for arts students my experience was that eighty percent of university students avoided science courses and twenty percent took them. Less than one percent was comfortable and did well in both. Interestingly, this percentage increased as more women moved in to sciences.

The challenge facing anyone trying to counter the exploiters is to bring logic, clarity and understanding in a way a majority of people can understand. You can write a book or make a movie that satisfies scientists, but a majority of the public will not understand. If you write for a wider audience, scientists will say it oversimplifies. Many have faced the challenge with documentaries and books about climate. Martin Durkin faced the challenge commendably with his documentary *The Great Global Warming Swindle*. A good book that straddled the dichotomy is Essex and McKitrick's *Taken by Storm (Revised edition)*, but many say they get lost.

It's a problem science books face even if they're tailored for the general market. How many people read and understood Stephen Hawking's *A Brief History of Time*? Yet it was a massive best seller.

It's a challenge *Scientific American* faced as a journal on science for general consumption. Scientists reading articles outside their discipline found them interesting, albeit arcane. With one in their discipline they realized it was oversimplified and inadequate. As a business and losing market share they decided to boost sales by becoming sensational, which included touting the false science of climate change.

A major challenge for education is to prepare people for the evolving scientific and technologically dominated world. Many universities have different combinations of 'required' courses.

These variably include a science credit for arts students, and humanities or social science courses for all students. All students need to understand science, but all science students must know the history of science, the social impacts and therefore the responsibility its practice requires.

In his 1961 retirement speech President Eisenhower anticipated the corruption of climate science of the last thirty years.

Yet, in holding scientific research and discovery in respect, as we should, we must also be alert to the equal and opposite danger that public policy could itself become the captive of a scientific-technological elite.

We can only achieve by overcoming the general public fear of science through education. Then they can spot exploitation like that practiced by the CRU, IPCC, and extreme environmentalists or at least understand what the few skeptics who refused to be silenced were saying. This book explains what has happened with climate science and what the skeptics are saying. It provides a chronology and significance of events, it examines the most significant issues, including limitations of the data, inadequacy of the computer models, lack of understanding of major astronomical, atmospheric, oceanic and terrestrial systems and shows how they were misused and manipulated.

Finally, it provides dramatic examples from specialists on how their portion of science was inaccurate and inappropriately used to distort climate science.

Chapter 2

The Basics

Starting from scratch, Alan Siddons will take the reader through several steps needed to understand the principles of thermal transfer and to illustrate that back-radiating trace gases cannot make the earth's surface warmer than solar energy makes it.

The Weakness of a Constant Irradiance Model
by Alan Siddons

IT'S IMPORTANT TO understand that radiant energy models don't deal with sun and earth conditions as they actually exist.

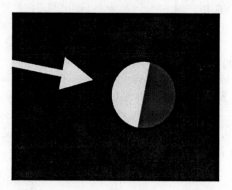

If a somewhat realistic model were used, the Earth would naturally be hottest at the noon equator, coldest at the poles, but beyond that, what? Wouldn't it be close to absolute zero on the shadow side?

Such a problem is hard to solve, especially considering that the earth also rotates, thereby adding the complication of exposure duration vs. heat-retention. Modelers therefore find it much easier to avoid these difficulties by imagining that sunlight has equal strength all over the planet. They do this by diminishing sunlight's power to a quarter of its actual value.

This way, the model has the same temperature everywhere—cooler than reality at the equator, warmer than reality at the poles, cooler than reality on the dayside and warmer at night. On this imaginary earth it's the same temperature everywhere.

This is why, instead of the 1368 watts per square meter that the real sun actually radiates toward the earth, most often you'll see it expressed as 342 watts. 342W is what a modeler takes as the energy impinging on every square meter of the planet all at once. **All at once.**

Keep that in mind. Like the summer sun in the Arctic, a modeler's sun never sets.

Forgetting this can lead to confusion. Just as the model sun always radiates 342 watts, a simplified model earth absorbing this amount emits 342 watts in return. Does this imply that the earth 'loses' 342? No, because it is constantly gaining 342 at the same time.

Emission can't occur without absorption. Effectively, a steady-state model makes the two identical—a simultaneous phenomenon. In particular fact, since there are some irradiance losses in real life, the earth model we go by continuously absorbs/emits around 240W rather than 342W.

One might picture this energy as water being pumped along a pipeline.

100% of the solar power that the earth absorbs is continuously emitted into the vacuum of space, only at infrared wavelengths.

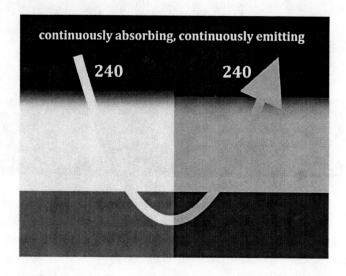

A modeler's planet earth, then, can never get colder than the heat it gains via 240 continuous watts per square meter. But can it get hotter?

Imagine that there's a kind of blockage 'up there'—such that solar energy enters but some of the terrestrial energy can't get out. Let's say it's 50%.

As you see, 120 terrestrial watts thus escape to space but the other 120 are blocked. Remember, though, the earth itself remains at 240 because the sun is always shining. The problem before us is to decide on the effect such a blockage might have.

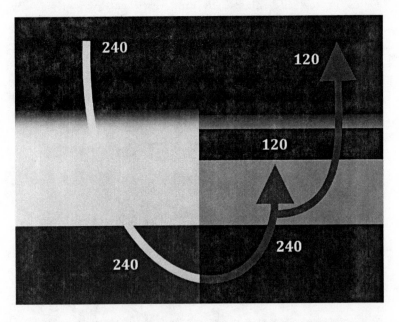

This problem is best approached by understanding why the sun is able to heat the earth in the first place. The profoundly simple answer is, because the earth is colder. It is less energetic than the sunlight that falls on it. Indeed, if the earth were a self-luminous body radiating the same 1368 watts per square meter the sun aims at it, nothing would happen. No heating would occur and there would be no transfer of energy. 1368 and 1368 would not combine to warm the earth a total of 2736 watts per square meter. To the contrary, if the difference between the sun's radiance and the earth's radiance were zero, the heating impact of the sun would be the same. **Zero.**

This may seem astonishing, but it's the nature of everyday reality.

For example, a spotlight cast on a dark object will brighten it, but if the object is glowing sufficiently on its own, there's no change of illumination—that is, no transfer of energy will result. The spotlight can't make the glowing object brighter because the spotlight is unable to add to the object's existing energy. There is no difference to overcome, and an energy transfer can only occur where a difference exists.

In short, radiant energy has but one way of exerting an effect: on a region of lesser energy. When a region possesses equal or greater energy, energy cannot flow there and cannot exert an effect. Greater thermal energy must move to lesser, hotter moving to colder.

This answers the question of a 50% radiant-blockage. The light cannot transfer its power downward—miraculously raising the earth to 360 W/m²—because the earth below has twice the energy. Without a difference to overcome, energy makes no difference.

Given a continuous heat input, then, no additional heating can occur by adding a radiant barrier, even if it blocks 100% of the outgoing energy. The pipeline analogy is apt: a cul-de-sac will merely stop the flow; it cannot amplify the amount of energy involved, i.e., it cannot raise the temperature. Otherwise, a beam of light could thermally excite a body to any magnitude; one watt per square meter could generate the heat of a billion watts or more. Just ensure that the target is surrounded by a reflector, and there is no limit to the power you'd obtain. You could melt an ingot of steel with a flashlight.

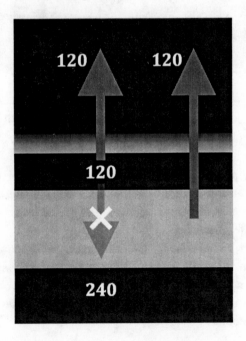

The Weakness of a Constant Irradiance Model

In reality, however, *the intensity of an object's emission is a signal of its temperature.* Sending that signal out and having it return does not change the signal. In other words, if the signal emitted by a hundred-degree body is directed back to it, the body 'reads' a hundred degree signal—and responds accordingly, i.e., its temperature remains the same. This is how the reflective coating in a thermos helps keep hot coffee hot. The light an object emits *is* a temperature signal. The reflective coating in a thermos serves to expose hot coffee to its own emission, which thereby sustains its temperature. Doubling-back the coffee's signal doesn't amplify the signal; it does not and cannot make the coffee hotter.

In sum, a constant-irradiance earth model is nothing but a constant temperature model. Although blocking its temperature signal (its emission) is widely believed to raise its temperature, this is not the case. A constant-irradiance model is thus unable to demonstrate the mechanism of a greenhouse effect, even though such a model (e.g., Kiehl-Trenberth) is always used to depict one.

A proper earth model would have to incorporate the factors cited earlier: intense sunlight on one side with none on the other, the rotation period, subsurface heat retention and rate of release…and so forth. Yet to be mentioned factors may also play a role.

Only then—by a process of elimination—could a valid case be made that the some other factor heats the earth. As it stands, the model we're using is insufficient.

Chapter 3

Basic Geometry
by Alan Siddons

THE FOLLOWING DESCRIBES the standard assumptions behind planetary temperature estimates. Whether those assumptions are valid is another question, and is dealt with elsewhere.

Alan Siddons

This rendering of a ball has a single source of light and I've made it slightly gibbous (more than half-lit to the eye) in order to emphasize its three-dimensionality. What's observable about this ball is elementary, but vital.

First of all, the light source can illuminate only half of the ball at any one time. Secondly, most light falling onto it falls obliquely, for only one point on the surface is perpendicular to the light source, thus receiving the maximum amount of energy.

A two-dimensional disk has four times less surface area than a sphere of the same diameter. Perpendicular to a beam of light, though, a disk's flat surface is able to absorb the full intensity. A hemisphere, by contrast, absorbs the same total amount but that amount is spread over a larger area, thus diluting it.

And to complicate matters further, Earth rotates and that's another issue the climate models can't deal with.

This has a direct impact on the temperature the two surfaces can reach. A blackbody temperature equation for the disk goes like this:

$$\text{Kelvin} = (P \div 5.67)^{0.25} \times 100$$

14

Or, if you prefer, in this alternate form:

$$\text{Kelvin} = \sqrt[4]{\frac{P}{5.67}} \times 100$$

where P is the power of the beam impinging on the disk. Let's call P 1000 W/m²...

Ergo,

$$(1000 \div 5.67)^{0.25} \times 100 = 364.42$$

So 364.42 Kelvin—or 91.27° Celsius—is the highest temperature the disk can reach. Notice another simple thing: the average and peak temperatures on the disk are identical, for the disk is receiving the same amount of energy everywhere.

The temperature equation for a sphere requires an adjustment. Since we know that the radiant power is diluted four times on account of its distribution over a greater surface area, we divide the initial 1000 W/m² by 4.

Ergo,

$$(250 \div 5.67)^{0.25} \times 100 = 257.69$$

So 257.69 Kelvin—or -15.46° Celsius—is the highest AVERAGE temperature the disk can reach. But in this case the average and the peak temperature are not the same, for the sphere is not receiving the same amount of energy everywhere.

It is crucial to understand this distinction. Only one point on a sphere faces radiant energy directly. For that reason, only this single point can reach the temperature of a perpendicular disk. There is a simple way to quantify that temperature. Once you have determined the sphere's average temperature in Kelvin, multiply it by the square root of two.

Ergo,

$$257.69 \times \sqrt{2} = 364.42$$

In other words, the sphere's peak temperature and the disk's temperature are identical.

Let's test these equations in a real-life application. We will adopt NASA's figure of a 1370 W/m² solar constant and have this fall onto the earth's moon, a sphere whose albedo (reflectance) is given as 0.07, thus an absorptance of 93%. So we divide radiance by four: 342.5, and multiply 342.5 by 0.93 to correct for reflection losses, obtaining 318.53 W/m².

Ergo,

$$\text{Average Kelvin} = (318.53 \div 5.67)^{0.25} \times 100 = 273.77$$

Now to determine the peak temperature on that spherical surface, multiply average Kelvin by the square root of two.

$$273.77 \times \sqrt{2} = 387.17\text{K}$$

Alternately, going for the peak temperature alone, 1370 W/m² × 0.93 = 1274.10 W/m² absorbed.

So:

$$\text{Disk Kelvin} = (1274.10 \div 5.67)^{0.25} \times 100 = 387.17\text{K}$$

The same.

Basic Geometry

In the exposition below, we quote NASA[1] on how they handle the problem:

> For slowly rotating planets like Mercury and the Moon, one must take into account that these bodies receive energy over their projected (disk) areas and emit energy, not over their full spherical surface areas but only over the same projected areas because the remaining surface area is considered to be too cold to radiate a significant amount of energy back to space. For such bodies, the thermal equilibrium is thus established when:

$$\sigma \varepsilon T_E^4 \times \pi R^2 = \frac{(1-A)L \times \pi R^2}{4\pi d^2} \quad (12)$$

> and

$$T_E = \sqrt[4]{\frac{(1-A)L}{4\pi\sigma\varepsilon d^2}} \quad (13a)$$

> or

$$T_E = 394\sqrt[4]{\frac{(1-A)}{\varepsilon d^2}} \quad (13b)$$

> where, as before, the Sun-Earth distance d is expressed in AU. A comparison of equations (10) and (13a) shows that for slowly rotating planets, the equilibrium temperature is higher by a factor equivalent to the fourth root of the projected area (i.e., the ratio of sphere surface-to-disk area),

---

[1] http://gltrs.grc.nasa.gov/reports/2001/TM-2001-210063.pdf

namely the fourth root of 4. Apply equation (13a), for $d = 1$ AU, $\varepsilon = 1$, and $A = 0.07$, to Earth's moon to obtain:

$$T_E = 394\sqrt[4]{\frac{(1-0.07)}{1^2}} = 387K \quad (13c)$$

which is the maximum temperature at the lunar equator at noon.

387K. NASA arrives at the same result. By the way, notice the 394 term above? That would be the peak Kelvin temperature if the 0.07 albedo loss weren't factored in, as so:

Disk Kelvin = $(1370 \div 5.67)^{0.25} \times 100 = 394.26K$

Divide that by the square root of two and you have the average temperature of a perfectly-absorbing sphere, 278.78 Kelvin.

Summary

The above outlines the assumed thermal response of a sphere that's absorbing radiant energy and verifies the standard methods with simplified blackbody equations.

Chapter 4

The Impact of an Atmosphere
by Alan Siddons

IF SCIENTISTS OF the past had known that the temperature of every planet with an atmosphere rises in direct proportion to atmospheric pressure, do you suppose they would have come up with a theory that attributed heating to the presence of certain trace gases that occupy less than one percent of our atmosphere?

No, of course they wouldn't have. Yet trace-gas heating theory has taken root so firmly by now that fresh perspectives have gone utterly ignored.

In Figure 1 you will find a graphic describing the altitude/temperature profile for Jupiter.

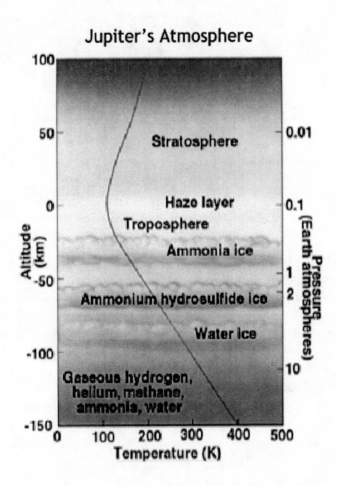

Figure 1: Jupiter's Atmosphere[1]

Notice that atmospheric heat rises with pressure. Is that the greenhouse effect at work?

[1] http://astronomy-guide.blogspot.com/2010/01/jupiters-layers-of-gas.html

In Figure 2 we see another view of Jupiter's temperature profile.

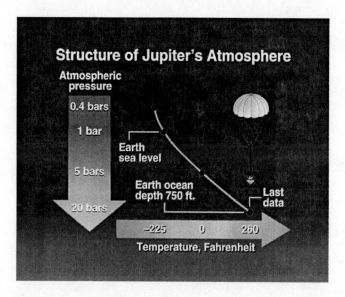

Figure 2: Structure of Jupiter's Atmosphere[2]

Atmospheric heat rises with pressure. Is that the greenhouse effect at work?

---

[2] http://www.solarviews.com/cap/craft/013sei.htm

In Figure 3 we see a graphic describing Saturn's altitude and temperature profile.

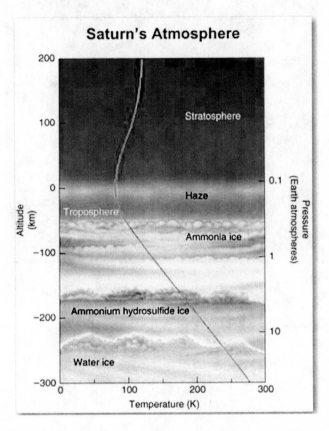

Figure 3: Structure of Saturn's Atmosphere[3]

Atmospheric heat rises with pressure. Is that the greenhouse effect at work?

3

http://physics.uoregon.edu/~jimbrau/BrauImNew/Chap12/FG12_0
4.jpg

In Figure 4 we see a graphic describing the altitude/temperature profiles of the four outer planets.

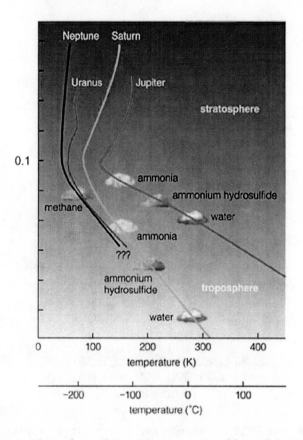

Figure 4: Altitude and Temperature Profiles of the Four Outer Planets.[4]

Atmospheric heat rises with pressure. Is that the greenhouse effect at work?

---

[4] http://astronomyonline.org/SolarSystem/JupiterIntroduction.asp

In Figure 5 we see a graphic describing the altitude/temperature profile for Venus.

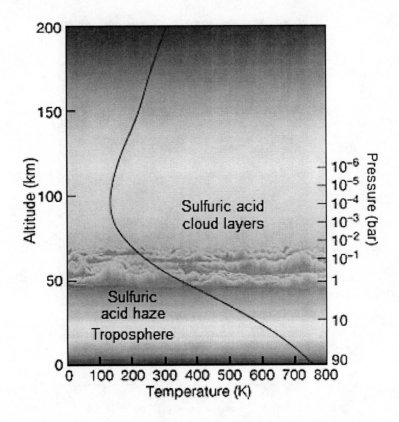

Figure 5: The Altitude and Temperature Profile of Venus[5]

Atmospheric heat rises with pressure. Is that the greenhouse effect at work?

---

[5] http://www.daviddarling.info/encyclopedia/V/Venusatmos.html

In Figure 6 we see a graphic showing the altitude/temperature profile for the earth.

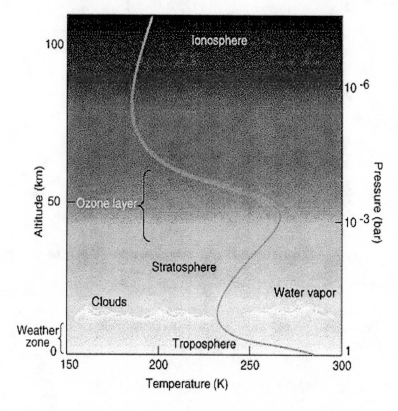

Figure 6: The Altitude and Temperature Profile of the Earth[6]

Atmospheric heat rises with pressure. Is that the greenhouse effect at work?

---

[6] http://www.astro.virginia.edu/class/oconnell/astr121/im/earth-atmprof-CM.jpg

To review:

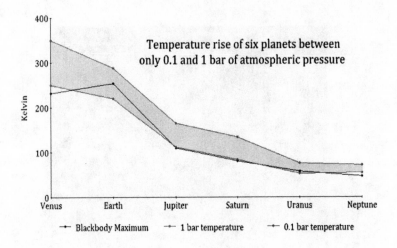

Figure 7: The Altitude and Temperature Profile of Six Planets

As these graphs indicate, between 0.1 and 1 bar of pressure, the atmospheric temperature of every planet rises above a predicted blackbody limit.

Is that the greenhouse effect at work?

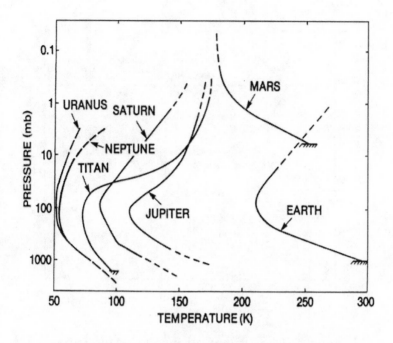

Figure 8: Pressure vs. Temperature of the Atmospheric Planets[7]

All planets with a substantial atmosphere show the same behavior, even Saturn's moon Titan. The atmosphere of Mars is just too vacuous to do the same. Once again, look at Jupiter's atmosphere, composed almost entirely of hydrogen and helium, which are not so-called 'greenhouse gases.'

Notice where the heating begins—like clockwork.

---

[7] http://lasp.colorado.edu/~bagenal/3720/CLASS14/AllPlanetsT.jpg

Alan Siddons

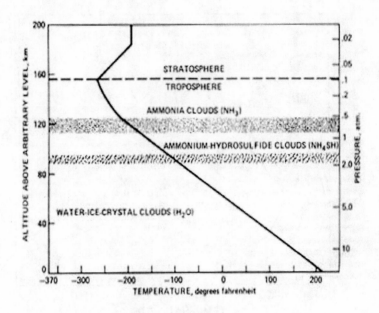

Figure 9: Constituents of Jupiter's Atmosphere[8]

Is this profile due to 'downwelling flux' from 'back-radiating' gases or simply due to the HEAT generated by mounting pressure?

The greenhouse effect theory was concocted to explain why the earth is warmer than Stefan-Boltzmann's Law predicts.

Yet, *every* planet is warmer than predicted.

Might something be seriously wrong with the prediction method?

---

[8] http://rst.gsfc.nasa.gov/Sect19/Sect19_15.html

Chapter 5

The Mother of all Averages
by Alan Siddons

Introduction

A BLACKBODY IS a theoretical entity that responds perfectly to radiant energy. Being perfectly absorptive ('black') to all frequencies of electromagnetic radiation, a blackbody heats up in a very predictable way. Measure for measure, a blackbody is the most thermally efficient object possible.

Now, if a blackbody were a planet it would take the form of a sphere. In radiative physics, a blackbody 'sphere' is effectively a flat disk that's been expanded four times and placed at twice the distance from the sun, thus allowing the inverse square law to reduce radiance on that disk by four times as well. Four times larger, but also four times less energized. Given 1 unit of irradiance on a disk, then, the same irradiance on a blackbody sphere equals 0.25 units. The question is: does this version of a sphere mimic a real one?

Theoretically, it would seem so.

29

A disk's surface area consists of its radius squared × π. So, with a radius of 1, a disk will have a surface area of π, 3.14. A sphere's greater surface area consists of its radius squared × π × 4, however. The same disk converted to a sphere will therefore have a surface area of 12.57, four times more than the original disk. Because this converted disk is four times larger but is exposed to the same amount of energy, each part receives four times less energy. It's the same effect as diluting whiskey with water.

To belabor the obvious even further, the earth's sphere is made up of two hemispheres, call them A and B. The sun illuminates A and leaves B in the dark. Since each hemisphere has twice the surface area of a disk, X watts per square meter directed at Hemisphere A gets diluted to 0.5X W/m² on its surface and of course 0 W/m² get spread over Hemisphere B. The average amount of light absorbed by A and B combined, then, is (0.5X + 0X) ÷ 2 = 0.25X.

In short, it always works out the same: a sphere absorbs four times less per surface area than a disk. Thus it seems reasonable to calculate temperature on this basis. Simply adjust radiance to 0.25, apply a radiance vs. temperature constant, and there you have your temperature. And, in fact, this is the accepted procedure.

There's a problem with this, however. And a huge one at that because radiance and temperature don't operate 1 to 1 together but on the basis of a 4th power law.

For example, if X watts of radiant energy raise an object's temperature to T (in Kelvin), then 16X is needed to raise the object to 2T. In other words, an object that has doubled its temperature is 16 times (2 to the 4th power) more energetic than before.

Because of this inequality between two quantities, 2 units of sunlight on surface A and 0 units of sunlight on surface B bring about two temperatures that are very different in combination than 1 unit of sunlight on both surfaces.

To prove this, let's do some calculations with real numbers.

• A blackbody disk exposed to 100 W/m² reaches a uniform temperature of 205K.

• Under the same circumstances a sphere supposedly absorbs four times less energy and reaches an average of 145K.

• But two hemispheres will reach 172K and 3K respectively (3K being the practical bottom limit in space), with an average of 87.5K, or 60% of the temperature predicted for a sphere.

For example, consider a planet that keeps one face to the sun. Half the planet's surface is constantly absorbing the available radiance while the other half absorbs nothing. Just as a perpendicular disk absorbs all the radiant energy impinging on it, a double-the-area hemisphere absorbs half, relatively speaking. As noted above, the result is 0.5 × radiance and 0 × radiance, yielding two temperatures to average: 172K and 3K in this case.

In terms of sunlight on a planet, then, the other hemisphere doesn't exist.

Thus, for a planet keeping one face to the sun, the traditional divide-by-four formula for temperature is inappropriate and misleading. The standard method robs Peter to pay Paul, underestimating the illuminated hemisphere's temperature for no good reason—while arbitrarily adding heat to the shadow side.

Yet, at any moment in time, every planet has but one face to the sun. Instant by instant, one hemisphere absorbs all the radiant energy available while the other absorbs none. No matter the scenario, nothing can alter the fact that one side is lit while the other side is in darkness. For decades this has been an unrecognized error in standard blackbody calculations for planets. An 'average radiance equals average temperature' assumption is clearly incorrect.

The hemispherical formula $(0.5X + 0) \div 2 = 0.25$ is a perfectly valid description of average radiance absorbed on a complete sphere. But this formula must be adhered to for determining temperature as well, $(T + 3) \div 2$, although the result is stunningly different from what people have been led to expect.

As one proof of the standard method's illegitimacy, notice that if you follow the divide-by-four formula, you cannot answer the simple question of how warm an illuminated hemisphere is. You have only an average spherical temperature to go by with no handle on any figures that comprise this supposed average.

Ramifications

Perhaps the first thing to point out about the geometrically justifiable rule of $(T + 3) \div 2$ is that it is most applicable to a sphere whose depth and conductivity may be regarded as 0. To understand this in converse terms, take a round pebble floating in outer space.

Exposed to 100 W/m^2, the pebble's outer surface will initially transfer warmth to its interior. In other words, the pebble will take time to warm up. Once conductive transfer has gone as far as it can go, there's no other means to store the heat, so the surface temperature will climb to a maximum—averaging 172K on the hemisphere facing the radiance.

But what of the other hemisphere? If the pebble is small enough, it's conceivable that nearly 100% of the pebble's acquired heat will migrate to the cold side, in which case both sides of the pebble will be at 172K, an average temperature 19% higher than predicted for a sphere absorbing 25% of the available radiance.

The larger the object, the less it can conductivity transfer heat to the cold side, but there's still stored heat to consider. If the sphere in question is a rotating planet and its soil holds onto 20K during the night, then the two sides will average $(172 + 20) \div 2$, i.e., 96K. The planet will be "hotter" than predicted by geometry, but due to nothing more than a surface possessing depth and not releasing its heat instantaneously.

Dividing a sphere's radiant energy by four is thus geometrically unjustified…a wild stab in the dark. Unless one knows how much heat the sphere can transfer internally and retain during rotation, there is no legitimate way to stipulate its average temperature.

A blackbody calculation is merely guesswork that an actual physical body is under no obligation to obey. Qua sphere, a body can reach a temperature of $(T + 3)/2$ all the way up to $(T + T)/2$, temperatures lower and higher than a simplistic divide-by-four formula.

As a corollary, these facts also demonstrate that there's no such thing as 'radiative equilibrium', i.e., no condition set by a vague calculation that forces a planet to adjust its temperature.

It is believed, for instance, that the earth's 'true' temperature is 255K, which would correspond to an ideal (blackbody) radiant emission of 240 W/m². It is further believed that an emission less than this must be compensated for by raising the temperature until the emission equals 240 W/m². Thus a radiative bottleneck is presumed to compromise the earth's emission such that an extra 150 W/m² are required to emit 240 W/m² in total. By this logic, the surface unaccountably rises to 288K, thus emitting 390 W/m² that get bottlenecked—but since 240 W/m² ultimately emerge, the 240 criterion is satisfied.

Yet nothing defines this criterion except a loosely-formulated temperature estimate that doesn't incorporate real conditions. A rational estimate must begin by assuming a half-lit and half-dark sphere and proceed from there.

The Mother of all Averages

In short, a planet's true temperature can only be guessed
at within a range of mathematically tenable possibilities,
beyond which actual empirical measurements are demanded.

Midpoint Conclusions

It has been demonstrated that the widely-accepted divide-by-
four rule cannot reliably predict the actual temperature
conditions on a globe due to the deviations inherent in a 4th
power law, which is also a 4th root law. To explain further,
sixteen times more energy brings about a doubling of
temperature because temperature conforms to the fourth root
of the radiant energy. Thus,

1 unit of radiance = $\sqrt[4]{1}$, i.e., one unit of temperature.

2 units of radiance = $\sqrt[4]{2}$, or 1.189207 units of
temperature.

4 units of radiance = $\sqrt[4]{4}$, or 1.414214 units of
temperature.

8 units of radiance = $\sqrt[4]{8}$, or 1.681793 units of
temperature.

16 units of radiance = $\sqrt[4]{16}$, or 2 units of temperature.

In detail, then, the divide-by-four practice consists of
mistakenly dividing a uniform disk temperature by the fourth
root of four.

Observe. A surface perpendicular to a radiant source of
1368 W/m² (the earth's solar constant) will reach a
maximum temperature of 394.11K, while a sphere under the
same conditions is believed to receive ¼ the energy because
of a four times greater surface area—and therefore reach a
temperature maximum of 278.68K.

In mathematical terms this means:

$$394.11 \div \sqrt[4]{4} = 278.68K$$

279K is traditionally cited as the earth's highest possible blackbody temperature[1].

But a hemisphere absorbs ½ the radiance available to a disk because its surface area is merely two times greater. The relationship between radiant energy and temperature therefore dictates that the hemisphere's average temperature is:

$$394.11 \div \sqrt[4]{2} = 331.41K$$

Given the other hemisphere absorbing zero, thus falling to 3K, the total sphere's temperature will average 167.20K.

Although real objects can reach temperatures very consistent with the Stefan-Boltzmann radiance vs. temperature formula, they take TIME to do so because their conductivity transfers heat internally. Until that heating process reaches saturation, a real object falls short of the predicted temperature. That's a key detail which the abstract physics of radiative forcing can't solve at a distance. You have to know the material's conductive properties.

Absent such specific information, the temperature estimate for a planet can only proceed on blackbody assumptions. Trimming the solar constant to average albedo, the angle of radiant energy on the planet's surface determines

---

[1] For further discussion, see *An Analysis and Procedure for Determining Space Environmental Sink Temperatures with Selected Computational Results.*

http://gltrs.grc.nasa.gov/reports/2001/TM-2001-210063.pdf

the temperature, followed by an equal allotment of 3K for the shadow side. The result will take a form like this:

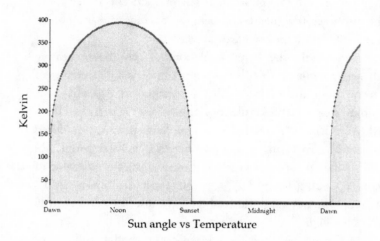

Sun angle vs Temperature

Here is a thermal profile of blackbody earth, for instance. The symbols denote dawn, noon, sunset, and midnight. With 1368 Watts per square meter on a spherical body that reflects no light, the peak temperature reaches 394K, and the lowest temperature 3K (or ideally 0). A full dawn-to-dawn cycle consisting of 360°, my spreadsheet reports the average temperature within that span as 170.14K. Since the angle of incidence is plotted only in 1° increments, the combined temperatures can be considered an estimate, but it's in good agreement with the geometric formula that says the average temperature is 167.20K.

Evidentiary Support

To review what we've seen up to now, the traditional method of dividing radiant energy by four to determine a planet's temperature neglects the fact that under real conditions the light-receiving hemisphere will reach a temperature higher than predicted for the sphere as a whole and the dark hemisphere's temperature will fall dramatically lower, both hemispheres together comprising an average that cannot be reconciled with the standard calculation.

This cracks the very foundations of greenhouse heating theory, for the earth's 'base' temperature is still a matter of fuzzy conjecture—it's still an unknown quantity. A geometrically justified rule for a sphere's average temperature is $(A + B)/2$, A and B being two hemispheres considered separately. Factoring in the object's heat transfer properties with rotation rate can produce a more accurate estimate, of course, but $(X + 3)/2$ is the most legitimate initial assumption, not a disk-derived temperature that's consistently 68% too high.

Adhering to this logic leads to a 'bullet and plain' temperature profile, the natural result of low to high temperatures brought about by a varying daytime solar angle and flat-line nighttime temperatures. Conductive transfer and heat retention will necessarily alter this profile, of course.

As supporting evidence for all of the above, I offer our *Greenhouse Effect on the Moon*[2] paper, wherein the same kind of thermal profile emerged, both theoretically and empirically.

NASA investigators followed the same procedure for projecting a moon temperature. The divide-by-four rule provides nothing specific, only a non-specific average.

So, NASA used the radiance vs. temperature formula itself and, as I have done, applied a sine or cosine rule to the angle of incident solar radiation in order to project a range of expected surface temperatures at various times of the day and compare this prediction to in situ measurements.

---

[2] http://www.tech-know.eu/uploads/Greenhouse_Effect_on_the_Moon.pdf

Their angle of incidence program gave them a profile that came close to reality. Even then, however, actual measurements differed from the prediction. Why? Because their program could not anticipate internal conductive transfer.

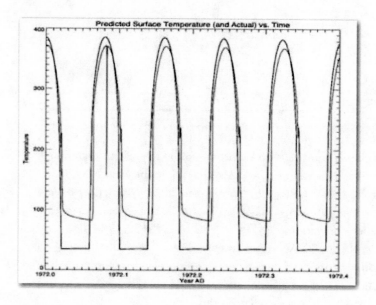

Blue is the profile predicted by the radiative knowns. The reason for nighttime temperatures not falling to 3K in this particular case is the earth's radiance during its 'full moon' phases.

For my own spreadsheet calculations, I plugged in a solar constant of 1368 W/m², an average absorption of 0.89 (1 minus albedo) and estimated that a 'full earth' at night would provide a 'floor' of 35K, all of which—in combination—gave me this temperature profile based on angle of incidence for sunlight.

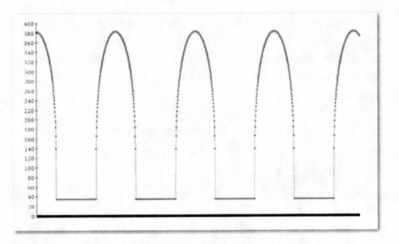

It closely mimics NASA's profile, although my predicted high is 382.79K whereas NASA's appears to be closer to 385K. From these inputs, the spreadsheet's AVG function over a 360 range returns 182.94K. This is in fair agreement with a paper calculation of $(321.89 + 35) \div 2 = 178.45K$.

NASA, however, assigns 274.5 for the moon, a temperature 96K higher! Indeed, simply divide radiance by four, correct for albedo, and you'll hit something close to that figure too. It's standard procedure.

As indicated above, conductive transfer during the warm-up cycle will bring surface temperature to a value lower than predicted until the transfer is complete, just as the reverse transfer of internal heat during the cool-down cycle will bring surface temperature to a value higher than predicted until it is complete. Referring to the NASA paper and the chart above, notice this is exactly what happens. The surface remains slightly cooler because conductive transfer never quite finishes. Some fraction of energy is still in the process of being tucked away when the bell rings and the sun passes its maximum height in the sky.

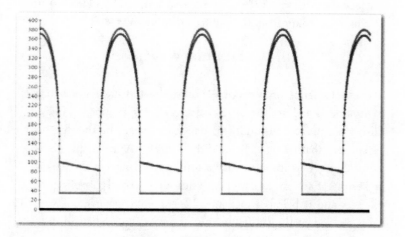

So too during the cool-down cycle: the surface temperature plummets to around 100K but then tends to hover there as the now-steeper thermal gradient between surface and depth draws out the internally-stored heat. Even this process never completes itself; however, as internal heat is still in the process of donating to the surface when it's saved by the bell and the sun begins to rise again!

On my spreadsheet, I duplicated the actual lunar profile to a fair degree, as depicted in blue. The average function that reported a theoretical temperature (red) as 183K reports 204K as the empirical temperature (blue), 70.5K less than NASA's 274.5K figure.

Conclusions

Even with the addition of a 'full earth', the moon is a model of radiative simplicity. If the standard method of estimating a sphere's temperature has any validity, it would certainly show in this case. But it doesn't. Real temperatures on the lunar surface deviate only 14% from those predicted by $(T + 35) \div 2$ while diverging from the published value by 74%[3].

Summary

Reduce radiance on a disk by four times and its temperature will indeed fall to the level calculated for a sphere. But a disk's temperature is uniform, which can never be the case on a sphere that is half lit and half dark. A real sphere is something very different than a four-times-larger disk. This is a serious flaw in radiative physics as currently applied to planets and it brings about an inherent 68% error compared to a geometrically justifiable rule of averaging sunlit and dark hemispheres as an initial guess.

It is alarming that the practice of using 25% irradiance to set a planet's temperature hasn't been noticed as a mistake before. More alarming still is that this erroneous formula has morphed into a 'law' of radiative equilibrium, the notion being that if a planet's temperature doesn't conform to a (flawed) calculation, so-called greenhouse gases are able to raise the planet's temperature until it does conform.

There is no physical reality behind a planetary blackbody estimate, thus no necessity driving a planet to adjust to it. Up to the present, climatology appears to have trusted a string of unexamined fictions—blackbody calculations being foremost among them. If climatology is to become a true science, these fictions have to be discarded and replaced with a strict regard for evidence.

---

[3] http://nssdc.gsfc.nasa.gov/planetary/factsheet/moonfact.html

Rediscovering RW Wood

You might have seen this passage several times but never noticed a telling detail before. It describes Professor R. W. Wood's greenhouse experiment in his own words[4].

> *To test the matter I constructed two enclosures of dead black cardboard, one covered with a glass plate, the other with a plate of rock-salt of equal thickness. The bulb of a thermometer was inserted in each enclosure and the whole packed in cotton, with the exception of the transparent plates which were exposed. When exposed to sunlight the temperature rose gradually to 65°C, the enclosure covered with the salt plate keeping a little ahead of the other, owing to the fact that it transmitted the longer waves from the Sun, which were stopped by the glass. In order to eliminate this action the sunlight was first passed through a glass plate.*
>
> *There was now scarcely a difference of one degree between the temperatures of the two enclosures. The maximum temperature reached was about 55°C.*

So, observe what Wood observed: his completely transparent salt enclosure reached a higher temperature than the infrared-opaque glass enclosure. Yet the IR opacity of glass is supposed to yield a higher interior temperature due to the blockage of outgoing heat rays. No indeed, however. Only after hobbling the salt pane with glass did the temperature of the two enclosures agree.

The sun radiates a range of wavelengths, including thermal infrared. But glass tends not to let infrared pass

---

[4] www.tech-know.eu/uploads/Note_on_the_Theory_of_the_Greenhouse.pdf

freely. Instead, glass absorbs and radiates it. Since the infrared-transparent pane—the salt plate—lets more sunlight into his box, then the interior got hotter than the glass-covered box.

The point is that IR-absorbing gases reduce the amount of radiation we receive from the sun. More than this, however, Wood's experiment showed that trapping heated air was the only factor involved—since an absorption-free scenario yielded the highest temperature. Yet the selective absorptivity of glass became the very basis for the atmospheric theory, which is demonstrated ad nauseam in the section below called *A Long List of Misconceptions*.

Comparing Apples to Pears

The thermal behavior of a real body vs. a blackbody can be compared to a race between the tortoise and the hare.

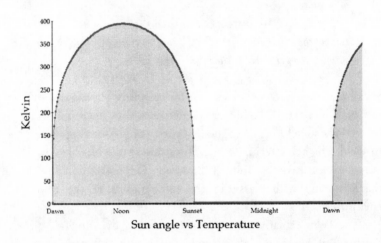

Sun angle vs Temperature

Constrained to emit 100% of the energy that impinges on it, a blackbody is unable to store any heat. As a result, the first slanted ray of light at dawn will raise its temperature immediately and predictably. The blackbody will thus reach its maximum temperature at solar noon, after which its temperature will fall as fast as it rose. Exposed to no light at night, the blackbody will radiate no energy at all—meaning that it's at absolute zero HALF of the time!

A real body is not as thermally receptive or responsive, however. It doesn't heat up as fast precisely because it's busy storing heat, conducting it internally into itself rather than fully radiating it. So it never gets as hot. But then it never gets as cold either. Reaching its highest temperature in the *afternoon*, it then begins to cool. And as it does so, the stored heat below now creeps toward the surface *because heat always flows from warmer to cooler*. In effect, a real body is a thermal battery—something that's especially handy in the dark. A blackbody has no such attributes.

This is roughly how such a difference might play out, with both bodies starting off at zero.

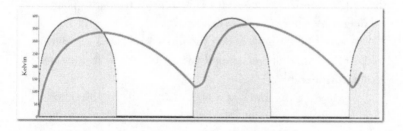

By the second dawn, the tortoise is ahead and its average temperature—with a lower high but a higher low—will thereafter keep exceeding the nimble hare's.

I should point out that a late-peaking phenomenon, a signal of the conductive storage of heat, is not just conjectural but is a matter of empirical fact.

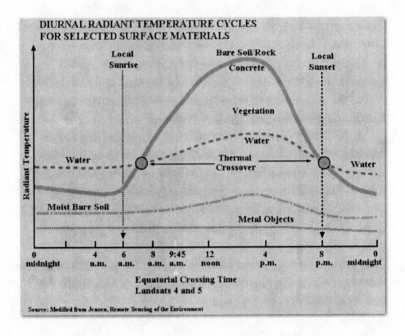

Indeed, this same phenomenon was also observed on our barren, waterless moon after Apollo astronauts planted temperature sensors on the surface[5].

A crucial difference is that the moon endures a two-week night rather than one of around twelve hours. So it does cool down considerably.

But still not as much as a maximally radiating blackbody. And this gives it a higher than predicted average temperature. The blue zone depicts the moon's thermal handicap, the orange its advantage.

---

[5] www.tech-know.eu/uploads/Greenhouse_Effect_on_the_Moon.pdf

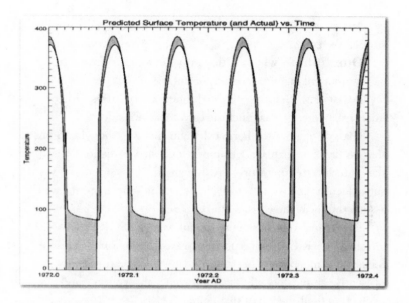

Moral of the story: A blackbody equation cannot predict a real body's temperature.

Yet the earth's 'base' temperature, the central premise of greenhouse theory, is calculated by a blackbody equation.

How Does Air Get Warm?

Deprived of a heated surface to make contact with, air could only be heated by radiative transfer, which would be unfortunate for gases that are transparent to radiation. In the real world, however, the atmosphere is not deprived of a heated surface to make contact with. Thus it gets heated directly, not radiatively. Considering, then, that CO_2 is only able to intercept about 8% of the earth's heat rays in the first place, and is outnumbered 2600 to 1, it's obvious that the majority gases excite trace gases far more than the other way around.

Moreover, 100% of this heated atmosphere is radiating IR toward the earth.

Question: So why is it that only radiation from the trace gas component is held to be important?

Answer: Because the founders of this theory misconstrued why glass enclosures get warm inside.

Glass is opaque to thermal-IR and this was thought to be the heating mechanism…trapping outward radiation raised the interior temperature. Although this assumption was subsequently proved wrong, the same mechanism was assumed to heat the earth's atmosphere.

By further misconstruing an infrared ABSORBER as an infrared BARRIER, IR-responsive trace gases became the sole focus of atmospheric heating. In short, climate science is presently mind-locked on infrared absorption and is neglecting the flip-side of that coin.

Absorption and Emission

Let's look at Kirchhoff's Laws.

Relative to the observer, an absorption spectrum signifies that a cooler gas is in front of a warmer (therefore brighter) body. This very fact alone proves that the cooler gas isn't heating the warmer body, i.e., the earth. I must say 'relative to the observer', of course, because from another angle of view, an observer will notice that the 'missing' wavelengths 'absorbed' by this cooler gas are radiating from it, creating an EMISSION spectrum.

In reality, no energy is trapped. What is being captured is simultaneously being released.

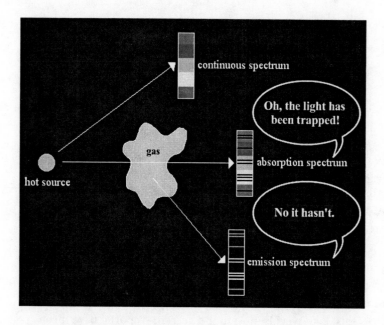

A previously-heated object will naturally cool down if left alone, isolated from any heat source. One cooling mechanism is of course radiation. In that sense, then, 'radiative cooling' is a legitimate concept, although it's a minor component compared to conductive and convective cooling.

This is why a spacecraft has such a hard time dumping internal heat to the surrounding vacuum of space[6]; radiative cooling is a sluggish process. But a constantly illuminated body that is radiating in response is not 'cooling down'. A simple thermometer will verify that. If this body is a blackbody, for example, its molecules are vibrating in 100% correspondence to the energy they're absorbing, and this vibration is CAUSING the electromagnetic energy they emit. Cutting off this outgoing energy, then, will not make the incoming energy vibrate those molecules vibrate any MORE.

---

[6]http://science.nasa.gov/science-news/science-at-nasa/2001/ast21mar_1/

This is why the suppression of 'radiative cooling' does not raise a body's temperature.

As I say, hot coffee in a thermos has a lot of lessons to teach.

Summary

In 1938, a play of H.G. Wells' novel *The War of the Worlds* was broadcast nationwide over the radio. According to news stories at the time, it led to instances of hysteria with many Americans believing that a Martian invasion was actually taking place.

What we have today in the global warming scare is a similar misapprehension, but, with a difference. Here, in effect, hoaxers warn that the alien onslaught is destroying more of the planet every year, while skeptics reassure people that this takeover is not to be feared; reports have been exaggerated. None but a few, however, are pointing out the simple fact that no Martians have landed at all.

The idea is that sun-heated ground and ocean emit infrared radiation into the sky, radiation that's absorbed by certain IR-sensitive gases (greenhouse gases) and emitted back to the ground, thereby raising the surface temperature and its consequent emission.

This bizarre mechanism is very clearly illustrated by the United Nations' Intergovernmental Panel on Climate Change.

5 Some of the infrared radiation is absorbed and re-emitted by the greenhouse gas molecules. The direct effect is the warming of the earth's surface and the troposphere.

Surface gains more heat and infrared radiation is emitted again

:he
... ... and is converted into heat causing the emission of longwave (infrared) radiation back to the atmosphere

But there are other explanations of the greenhouse effect. One is: decreasing a body's outgoing radiation increases that body's temperature, as if the two are on a see-saw. This is skewed causality, however. Reducing a body's temperature decreases its outgoing radiation, yes. But decrease its radiation per se, i.e., block it, and it simply stays at that temperature, like coffee in a thermos.

Decreased outgoing radiation leading to increased temperature is yet another theorem of greenhouse physics that has no place in real physics. You won't find a formula for it anywhere.

So there are at least two versions of the greenhouse heating effect: back-radiation vs. reduced radiative cooling, neither of which has evidence to support it. Reflecting an emitter's own radiation back to it doesn't raise its temperature in real life, only on paper. And, as nearly as empirical measurements can establish, the earth emits to space the same magnitude of radiation as it receives from the sun. Since there's no sign that any radiation is being blocked, then, the argument that the earth is 'cooling less' than it would otherwise dissolves.

By the way, if the UN's depiction of a magical heat-magnifying mechanism isn't enough to make you laugh, Dr. Michael Pidwirny, who runs Physical Geography.net[7], brings it into closer focus...

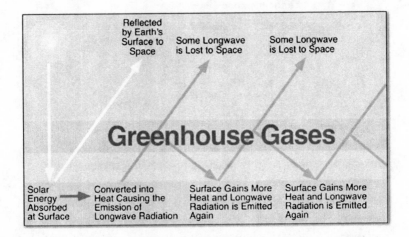

As you see, with greenhouse physics, anything goes. Once you decide thermal energy can be counted multiple times, you can get any temperature you want.

For more than a century now, the theory of an atmospheric greenhouse-effect gained ground only because academic eggheads lost contact with reality, having never grasped basic physics.

---

[7] www.physicalgeography.net/home.html

Chapter 6

Learning by Candlelight
by Alan Siddons

To see a world in a grain of sand...
—William Blake

WHILE ENJOYING A recent effect of global warming, a week-long blackout brought on by a freak ice-storm which devastated the central Massachusetts region, I had ample opportunity to contemplate how a candle's flame behaves.

It's often said that here on the earth's surface, air convection is the ruling heat-loss mechanism. And how. We're like fish living at the bottom of an ocean, yet are seldom aware of how our effort to generate heat is constantly thwarted by the very medium we're breathing. It's not that air is a good conductor; it's that once it does conduct it won't stand still. Due to gravity, heated air becomes lighter in weight and rises away, while cooler air is displaced downward and steals more heat from the source. This process shapes a candle's flame and even influences its color.

Hold a candle at any angle and the flame always points upward—away from the earth's center. The flame responds to gravity. It would otherwise look like a ball, not a teardrop, but the currents it generates push colder air into it, thus squeezing it into something more cylindrical. This air infiltrates the flame itself, so, although currents keep bringing in fresh oxygen to use, the cooling effect is profound. The net result is a vigorous flame that's too cool to burn efficiently. The black soot a candle emits is unburned carbon, a symptom of incomplete combustion. Due to air convection, then, a candle flame is never as hot as it *could* be—although it's brighter than it *would* be. All because air moves nimbly in a gravitational field.

The oddness of this being so familiar to us, the appearance of a candle in zero gravity is somewhat startling.

The flame is spherical because no convection occurs. It's blue because of complete combustion. It's dimmer because of a slower rate of oxygen replenishment in static air.

Learning by Candlelight

As I waited night after night for the electricity to return, candlelight kept teaching me about moving air's talent for removing heat, hampering any effort to keep warmth 'down here' by constantly sending it up and away. Good thing for a heat-containing roof, then; it lessens the harm considerably. The earth itself lacks any such roof, however. And imagining that certain radiation-absorbing gases provide one is only to confuse radiation with convection.

A physical lid over a heat source decreases the zone of circulating air, thus reducing the cooling rate. But an open 'lid' of gas that's capable of absorbing radiant energy will convect around like any other gas, stealing heat and doing nothing else except radiating the very energy it has received by radiation, having zero power to confine it.

Rather than limiting the area in which heat-loss occurs, then, a radiant absorber constitutes no barrier to radiation at all—it's merely a second radiator that relays heat away. And, just as there's no such thing as 'back-convection'—where a flame makes itself hotter by the air currents it creates—or 'back-conduction'—where a colder object raises the temperature of what it's in contact with—there's no such thing as 'back-radiation'.

Redirecting radiant energy back to the source cannot increase its temperature.

In all of its forms, heat spontaneously moves from a more intense zone to a lesser. What makes convection particularly dynamic and meddlesome is that a cool mass also keeps moving to the heat source—a double whammy.

A lot can still be learned by candlelight.

Chapter 7

What is an Average Temperature?
by Alan Siddons

Introduction: The Two Floors Problem

SAY YOUR HOUSE has two floors. Downstairs the temperature is at 72°, upstairs at 76°. You might conclude, then, that the house's average interior temperature is 74. But wait. Now you recall that the upstairs is 15% smaller. So should the average temperature be estimated thus?

Whole house $= 100\%$
Downstairs $= X$
Upstairs $= X \times 0.85$
Therefore $X + 0.85\,X = 100$, meaning that:
$1.85\,X = 100$ So $X = 54.054$

Thus:
Downstairs $= 54.054\%$ of the house
Upstairs $= 45.946\%$ of the house
'Weighting' the two temperatures, then…
Downstairs $= 72 \times 0.54054 = 38.91888$
Upstairs $= 76 \times 0.45946 = 34.91896$

Adding these two numbers, the house's actual average is therefore closer to 73.84°. Volume or area must always be factored in.

See how complicated an 'average temperature' can be? And you haven't even counted the crawl space in the attic! Finding an average temperature is more difficult with the Stefan-Boltzmann equation[1], however, because there's a 4th root involved.

Complication #1: The 4th Root Problem

Imagine two spots on a blackbody sphere are exposed to 50 and 100 watts per square meter. (Due to curvature, remember, a single light source gets spread out and becomes weaker.) Using the Stefan-Boltzmann equation, the two temperatures will be about 172 and 205 Kelvin respectively, i.e., an average of 188.5K. But the average irradiance is 75 W/m², which corresponds to 191K. That's 2.5 degrees off the mark. In other words, average temperature does not agree with average irradiance, and vice versa.

Take three spots at 100, 200, and 300 W/m². The average of course is 200 W/m². The temperatures are 205, 244, and 270 respectively, averaging about 240K. But 200 W/m², the average, equals 244K. Now you're four degrees off the mark. And so on, as you proceed to compare irradiance with temperature on each and every angle of a half-lit sphere. It's a huge problem to tackle.

Throw in rotation (i.e., the irradiance is constantly changing) and the heat-retention of various three-dimensional substances, and the problem runs out of control.

---

[1] http://hyperphysics.phy-astr.gsu.edu/hbase/thermo/stefan.html

Complication #2: The Minus 18 Degrees Problem

As for the famous minus 18° C surface temperature the earth is supposed to have without the greenhouse effect, that figure assumes a blackbody surface absorbing about 239 W/m² 'on average.' But check the Kiehl-Trenberth chart[2]. Due to clouds and other obscuring factors, the actual surface average is given as only 168 W/m². That figure corresponds to -40°C on the surface, meaning that it has to rise by 55 degrees, not 33, in order to reach the accepted average of +15. Anyone who tells you, then, that the 'greenhouse effect' makes the earth's surface 33 degrees warmer is merely confessing his (or her) own ignorance.

Conclusion: The Average Temperature without Temperatures Problem

Ask yourself what kind of 'average temperature' consists of no highs and lows and in-betweens? The earth's purported average temperature from Stefan-Boltzmann lacks any specifics—no information about average polar vs. equatorial differences and no information even about average day and night differences. What sort of average is that? This is why NASA engineers couldn't find any use for it[3]. And, as I say, it's because it really isn't an average temperature in the first place, it's merely the result of dividing irradiance by four and thoughtlessly parroting what an equation says.

---

[2] http://www.cgd.ucar.edu/cas/abstracts/files/kevin1997_1.html
[3] http://www.tech-know.eu/uploads/Greenhouse_Effect_on_the_Moon.pdf

Chapter 8

A Long List of Misconceptions
by Alan Siddons

WHAT ANIMATES GLOBAL warming concerns more than anything is the imaginary greenhouse effect and an equally imaginary law of physics called 'radiative equilibrium'. Energy out must equal energy in, this 'law' says. This sounds plausible on the face of it. In this view, if the light emitted by a heated object is suppressed in some way, its radiant energy will increase past the level of radiant input until it breaks through the barrier...in obedience to this 'law'.

This notion originates from a long-ago misconception about how glass greenhouses work, thus the family name this 'effect' goes by. It was believed that glass blocked the passage of 'dark radiation' (infrared) and kept storing energetic photons inside it. Once those photons accumulated enough power to overcome the glass barrier, radiative equilibrium was achieved.

So this is the scenario: sunlight enters, heat is generated and dark light is emitted. This dark light is amplified because of the blockage and finally exits at the same magnitude as the entering sunlight. But only after the light 'trapped' inside has raised the greenhouse's temperature. Since the barrier will keep raising the temperature until the barrier is broken, increasing the barrier's strength will get you any amount of internal heat you want.

If only that were true...

It is 19th century poppycock. And here's a telltale sign of it: why do you always see a 'layer of greenhouse gases' depicted overhead in illustrations about the 'greenhouse effect', when in fact these molecules are at their densest concentration right at your feet?

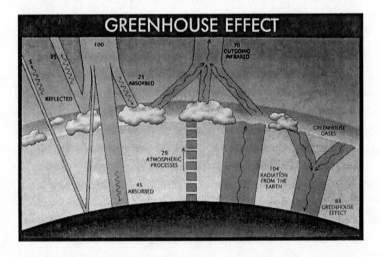

Because what these illustrations are showing you is the theory's genetic lineage. That 'layer of greenhouse gases' is merely a pane of greenhouse glass in another guise. There is no such 'layer'.

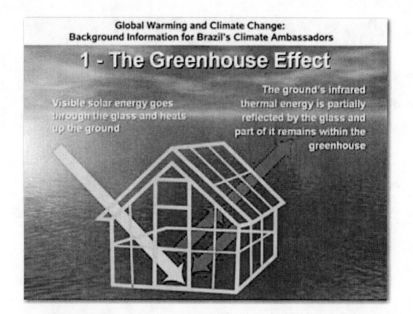

Global Warming and Climate Change:
Background Information for Brazil's Climate Ambassadors

1 - The Greenhouse Effect

Visible solar energy goes through the glass and heats up the ground

The ground's infrared thermal energy is partially reflected by the glass and part of it remains within the greenhouse

It started from a misconception about glass nearly 200 years ago—and it stayed that way.

In reality, greenhouses merely suppress convective heat-loss, preventing the heated air from dissipating. It is air that's trapped, not radiation; glass's response to infrared (IR) has nothing to do with it. Clear plastic bags will do just as well or even panes of polished salt crystals, which don't absorb IR at all. This is why salt crystals are used as windows in laboratory IR spectroscopy. Also, any infrared radiation absorbed by the glass is immediately re-radiated (scattered in all directions) by that glass—it does not constitute a radiative barrier.

**Thermal IR Image of a House, Showing IR Radiation
Passing Through the Glass Windows.**

This misconception is most famously known as the 'settled
science'. Although nothing is further from the truth, take a
look at what the messengers say.

PLEASE NOTE:
**None of what is described below actually occurs in
reality.**

—The Geological Society of London
The Greenhouse Effect arises because certain gases (the so-
called greenhouse gases) in the atmosphere absorb the long
wavelength infrared radiation emitted by the Earth's surface
and re-radiate it, so warming the atmosphere. This natural
effect keeps our atmosphere some 30"C warmer than it would

be without those gases. Increasing the concentration of such gases will increase the effect (i.e. warm the atmosphere more).[1]

—Bigelow Laboratory for Ocean Sciences
In a greenhouse, visible light (e.g., from the Sun) easily penetrates glass or plastic walls, but heat (in the form of infrared radiation) does not. The greenhouse effect refers to the physical process by which atmospheric gases allow sunlight to pass through but absorb infrared radiation thus acting like a blanket trapping heat.[2]

—The U.S. government's Environmental Protection Agency
The energy that is absorbed is converted in part to heat energy that is re-radiated back into the atmosphere. Heat energy waves are not visible, and are generally in the infrared (long-wavelength) portion of the spectrum compared to visible light. Physical laws show that atmospheric constituents—notably water vapor and carbon dioxide gas—that are transparent to visible light are not transparent to heat waves. Hence, re-radiated energy in the infrared portion of the spectrum is trapped within the atmosphere, keeping the surface temperature warm. This phenomenon is called the "greenhouse effect" because it is exactly the same principle that heats a greenhouse.[3]

[1] http://www.geolsoc.org.uk/webdav/site/GSL/groups/ourviews_edit/public/Climate%20change%20-%20evidence%20from%20the%20geological%20record.pdf

[2] http://www.bigelow.org/virtual/handson/greenhouse_make.html

[3] http://www.epa.gov/ne/students/pdfs/activ13.pdf

—Fort Lewis College, Colorado

This partial trapping of solar radiation is known as the greenhouse effect. The name comes from the fact that a very similar process operates in a greenhouse. Sunlight passes relatively unhindered through glass panes, but much of the infrared radiation reemitted by the plants is blocked by the glass and cannot get out. Consequently, the interior of the greenhouse heats up, and flowers, fruits, and vegetables can grow even on cold wintry days.[4]

—PlanetConnecticut.org

Glass is transparent to sunlight, but is effectively opaque to infrared radiation.

Therefore, the glass warms up when it absorbs some of the infrared radiation that is radiated by the ground, water, and biomass. The glass will then re-radiate this heat as infrared radiation, some to the outside and some back into the greenhouse. The energy radiated back into the greenhouse causes the inside of the greenhouse to heat up.[5]

—United Nations Framework Convention on Climate Change

Greenhouse gases make up only about 1 per cent of the atmosphere, but they act like a blanket around the earth, or like the glass roof of a greenhouse—they trap heat and keep the planet some 30 degrees C warmer than it would be otherwise.[6]

---

[4] http://physics.fortlewis.edu/Astronomy/astronomy%20today/CHAISSON/AT30 7/HTML/AT30702.HTM

[5] http://www.planetconnecticut.org/teachersadministrators/pdfs/lesson1.pdf

[6] http://unfccc.int/essential_background/feeling_the_heat/items/2903.php

—NASA

The "greenhouse effect" is the warming of climate that results when the atmosphere traps heat radiating from Earth toward space. Certain gases in the atmosphere resemble glass in a greenhouse, allowing sunlight to pass into the 'greenhouse', but blocking Earth's heat from escaping into space.[7]

—NASA

Why is this process called "The Greenhouse Effect?" Because the same process keeps glass-covered greenhouses warm. The Sun heats the ground and greenery inside the greenhouse, but the glass absorbs the re-radiated infra-red and returns some of it to the inside.[8]

—NASA

A real greenhouse is made of glass, which lets visible sunlight through from the outside. This light gets absorbed by all the materials inside, and the warmed surfaces radiate infrared light, sometimes called "heat rays", back. But the glass, although transparent to visible light, acts as a partial barrier to the infrared light. So some of this infrared radiation, or heat, gets trapped inside.[9]

—Deptartment of Atmospheric and Oceanic Science at the University of Maryland

A real greenhouse is enclosed by glass walls and ceilings. Glass is highly transparent in the visible wavelengths of the sun, so sunlight freely passes into the greenhouse. However, glass is highly absorbing in the infrared wavelengths characteristic of

---

[7] http://www.gsfc.nasa.gov/gsfc/service/gallery/fact_sheets/earthsci/green.htm

[8] http://www-istp.gsfc.nasa.gov/stargaze/Lsun1lit.htm

[9] http://www-airs.jpl.nasa.gov/News/Features/FeaturesClimateChange/Greenhous eEffect/

emission by earth's surface. Therefore, the infrared radiation emitted by the surface is efficiently absorbed by the glass walls and ceiling, and these surfaces, in turn, radiate energy back into the interior of the greenhouse, as well as outward to the environment. But clearly, a large portion of the infrared radiation from the surface does not pass outward from the greenhouse, and the equivalent energy is contained within the greenhouse environment.[10]

—How Stuff Works
Light passes through the glass into the greenhouse and heats things up inside the greenhouse. The glass is then opaque to the infrared energy these heated things are emitting, so the heat is trapped inside the greenhouse.[11]

—Enviropedia.org.uk
Greenhouse gases like water vapour, carbon dioxide, methane and nitrous oxide trap the infrared radiation released by the Earth's surface. The atmosphere acts like the glass in a greenhouse, allowing much of the shortwave solar radiation to travel through unimpeded, but trapping a lot of the longwave heat energy trying to escape back to space. This process makes the temperature rise in the atmosphere just as it does in the greenhouse. This is the Earth's natural greenhouse effect and keeps the Earth 33°C warmer than it would be without an atmosphere, at an average 15°C. In contrast, the moon, which has no atmosphere, has an average surface temperature of -18°C.[12]

---

[10] http://www.atmos.umd.edu/~owen/CHPI/IMAGES/grnhs1.html

[11] http://home.howstuffworks.com/question238.htm

[12] http://www.enviropedia.org.uk/Climate_Change/Greenhouse_Effect.php

—Physics Department, University of Alaska-Fairbanks

Greenhouse gases act as a blanket. Some of you may wonder how a green house takes solar energy and turns it into thermal energy. A good example of this is something you can observe every day in the summer in your own car. It happens when you leave you car in a sunny parking lot with the windows up. The solar energy is passing through the glass and is heating the cars interior. What's really happening is the short wave infrared waves are going in and are turning into long wave infrared waves, which cannot escape.[13]

—Climate.org

Fortunately, much of this infrared radiation is absorbed in the atmosphere by the so-called greenhouse gases, making the world much warmer than it would be without them. These gases act rather like the glass in a greenhouse, which allows sunlight to enter, provides shelter from the wind and prevents most of the infrared energy from escaping, keeping the temperature warm.[14]

State of Utah Office of Education

On a global scale, carbon dioxide, water vapor, and other gases present in the atmosphere are similar to the glass in a greenhouse. Ultraviolet radiation from the sun (having a short wavelength) can pass through the glass. Once inside the greenhouse, the ultraviolet radiation is absorbed by soils, plants, and other objects. Upon absorption, it becomes infrared radiation or heat energy having a shorter wavelength. Because of this, infrared radiation cannot escape through the windows. The windows act like a large blanket in which they

---

[13] http://ffden-2.phys.uaf.edu/102spring2002_Web_projects/C.Levit/web%20page. html

[14] http://www.climate.org.ua/ghg/ghgeffect.html

reradiate the infrared energy back into the greenhouse. This phenomenon naturally causes the overall temperature within the greenhouse to increase.[15]

—G.H.P. Dharmaratna, Director General Department of Meteorology

In order to understand the greenhouse effect on earth a good place to start is in a greenhouse. A greenhouse is kept warm because energy coming in from the sun (in the form of visible sunlight) is able to pass easily through the glass of the greenhouse and heat the soil and plants inside. But energy which is emitted from the soil and plants is in the form of invisible infrared radiation; this is not able to pass as easily through the glass of the greenhouse. Some of the infrared heat energy is trapped inside; this is the main reason why a greenhouse is warmer than outside.[16]

—Weather-Climate.org

This warming effect is called the "greenhouse effect" because it is the same process as that which occurs in a greenhouse on a sunny day. The glass is transparent to short-wave radiation but absorbs the outgoing long-wave radiation, causing a rise in temperature inside the greenhouse.[17]

—Eduhistory.com

The glass used for a greenhouse acts as a selective transmission medium for different spectral frequencies, and its effect is to trap energy within the greenhouse, which heats both the plants and the ground inside it. This warms the air near the

---

[15] http://www.usoe.k12.ut.us/curr/science/core/earth/sciber9/Stand_6/html/1e.htm

[16] http://www.lankajalani.org/Publications/Paper%20-%20Impacts%20of%20Climate%20Change.doc

[17] http://www.weather-climate.org.uk/04.php

ground, and this air is prevented from rising and flowing away. This can be demonstrated by opening a small window near the roof of a greenhouse: the temperature drops considerably. Greenhouses thus work by trapping electromagnetic radiation and preventing convection.[18]

—Northwestern University, Evanston, Illinois
Overview: Carbon Dioxide is identified as 'greenhouse gas' because of its ability to trap heat within earth's environment. Explain that the greenhouse effect works in a somewhat similar—but not entirely the same—way (see teacher notes and background supplement sheet for more information). The sun's rays pass through the atmosphere and warm the surface. The earth emits some of this energy back into space (like heat from a campfire). But gases such as carbon dioxide and water vapor (in clouds) absorb much of this energy and send it back to earth. People have come to call this process the "greenhouse effect" because it reminds them of how actual greenhouses, which are made out of glass and grow plants, let the sun's rays in while trapping much of the radiation that would otherwise escape.[19]

Remember:
None of the above actually occurs.
Yet authorities attest to it...

[18] http://www.eduhistory.com/greenhouse.htm
[19] http://www.letus.northwestern.edu/projects/gw/pdf/C09.pdf

Chapter 9

Examining the Greenhouse Theory
by Alan Siddons

Here's a chart that attempts to explain how radiative forcing works.[1]

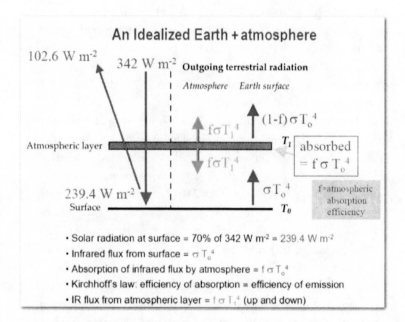

An Idealized Earth + atmosphere

102.6 W m^{-2}

342 W m^{-2} Outgoing terrestrial radiation

Atmosphere   Earth surface

$(1-f)\sigma T_o^4$

$f\sigma T_1^4$

Atmospheric layer

$f\sigma T_1^4$

T_1 absorbed $= f\sigma T_o^4$

239.4 W m^{-2}

Surface

σT_o^4 T_0

f=atmospheric absorption efficiency

- Solar radiation at surface = 70% of 342 W m^{-2} = 239.4 W m^{-2}
- Infrared flux from surface = σT_o^4
- Absorption of infrared flux by atmosphere = $f\sigma T_o^4$
- Kirchhoff's law: efficiency of absorption = efficiency of emission
- IR flux from atmospheric layer = $f\sigma T_1^4$ (up and down)

---

[1] http://courses.washington.edu/pcc588/lectures_2008/588_lect_010708.pdf

This comes from a university course but it is not unique. Gavin Schmidt of NASA uses the same example to explain the greenhouse effect to his readers.

To the left of the dotted line is incoming solar radiation. Of an initial 342 watts per square meter beamed at the earth, 102.6 are rejected and sent back to space, resulting in the earth absorbing 239.4 watts per square meter. To the right is outgoing—er, and also incoming—terrestrial radiation. It is understood that outgoing terrestrial energy must equal incoming.

The atmospheric function f plays a key role here. It represents the atmosphere's efficiency at intercepting terrestrial emission. As f rises, direct terrestrial emission to space necessarily declines. But, since atmospheric absorption increases as direct surface emission decreases, it performs the job of radiating the difference. This ensures that energy out equals energy in. On the other hand, what the atmosphere has absorbed from the surface also gets emitted back to the surface.

For instance, call terrestrial emission 240 instead of 239.4 and picture a 50/50 scenario.

The surface will emit only 120 W/m² to space because half is caught by the atmosphere. The atmosphere emits the 120 it has absorbed, bringing the earth to 'radiative equilibrium'. But that 120 is also radiated down to the surface, raising surface energy to 240 plus 120, i.e., 360 watts per square meter, quite a bit warmer now. A little more tweaking and you can get the surface to the requisite 390 W/m², enough to bring the earth's average temperature to 15° Celsius.

If people are gullible enough to believe such a scenario, and apparently millions do, they deserve what's coming down the road at them. Yet, this is what even many climate skeptics call 'the basic science'.

Substitute an infrared filter for that layer of 'greenhouse gases'. Like the gases, the filter is also transparent to visible light, but largely opaque to infrared. Direct a radiant heater at an infrared filter. According to greenhouse physics, you now have the equivalent of two radiant heaters because the infrared filter will absorb, say, 500 W/m² from the heater and emit that to the surroundings but also radiate 500 W/m² in the other direction, back to the heater. You get 1000 watts per square meter in all. Two heaters for the price of one.

But that's not all. Remember, the radiant heater will be heated by its own re-directed energy and thereby emit even more energy—which the glass will absorb and double, which will heat the heater more...

It's not only a perpetual motion machine—it accelerates to boot!

That such a childish fantasy threatens to destroy western civilization is incredible, but that's exactly the case.

More Discussion
By Heinz Thieme

The relationship between so-called greenhouse gases and atmospheric temperature is not yet well understood. So far, climatologists have hardly participated in serious scientific discussion of the basic energetic mechanisms of the atmosphere. Some of them, however, appear to be starting to realize that their greenhouse paradigm is fundamentally flawed, and already preparing to withdraw their theories about the climatic effects of CO_2 and other trace gases. [...] This is no surprise, because in fact there is no such thing as the greenhouse effect: it is an impossibility.[2]

---

[2] http://realplanet.eu/error.htm

And Again, More Discussion
By Tom Kondis

To support their argument, advocates of man-made global warming have intermingled elements of greenhouse activity and infrared absorption to promote the image that carbon dioxide traps heat near earth's surface like molecular greenhouses insulating our atmosphere. Their imagery, however, is seriously flawed.

A greenhouse is simply a physical structure that traps heated air. Solar radiation initiates the heating sequence inside a greenhouse when photons in the visible region of the electromagnetic spectrum, entering through glass or transparent plastic panels, are absorbed by surfaces of opaque objects. Reflected photons exit freely; neither they, nor their 'heat', are trapped inside. Drivers who regularly park their mobile greenhouses in sunny locations exploit this principle by placing reflective white cardboard behind their windshields to expel some before they're absorbed.

The second law of thermodynamics prohibits carbon dioxide from arresting or reversing the spontaneous downhill flow of energy, putting advocates in the awkward position of insisting that a trace atmospheric component's innocent participation in a natural heat dissipation process is responsible for warming a planet. The fictitious 'trapped heat' property, which they aggressively promote with a dishonest 'greenhouse gas' metaphor, is based on their misrepresentation of natural absorption and emission energy transfer processes and disregard of two fundamental laws of physics. Their promotional embellishments have also corrupted the meaning of 'greenhouse effect', a term originally relating the loose confinement of warm nighttime air near ground level by cloud cover, to hot air trapped inside a greenhouse.[3]

Further Greenhouse Comments

One thing that passes unnoticed by many greenhouse advocates is that water vapor plays quite a role in keeping the planet cool by absorbing incoming radiant energy. The blue line below is what we'd get without an atmosphere, the yellow line what we get with it. Sunlit temperatures on the earth's surface are appreciably less than those on our neighbor the moon, because our atmosphere intercepts incoming radiation. Given the conservation of energy law, 'greenhouse gases' cannot add heat to the earth's surface, but they can certainly reduce it.

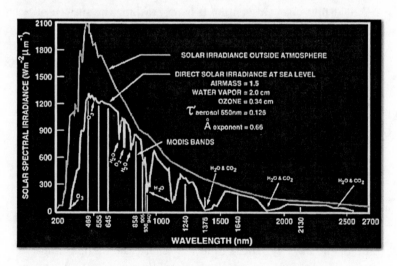

So too, consider this aspect of radiant-interception. The stratosphere, the layer of atmosphere just above the one in

---

[3] http://junkscience.com/Greenhouse/Kondis-Greenhouse.html

which we live, contains a thin layer of ozone (O_3). This layer wouldn't exist without the sun. Ozone is made of three atoms of oxygen. It's not a very stable molecule and it takes a lot of power to create it. When UV light hits a molecule of oxygen (O_2), it splits it into two atoms of oxygen (O). When one of these atoms comes into contact with a molecule of oxygen, they combine to make ozone.

The process also works in reverse—when UV light hits ozone, it splits it into a molecule of oxygen and an atom of oxygen.

Oxygen molecule + light = two atoms of oxygen.
Oxygen atom + oxygen molecule = ozone molecule.

This process is called the ozone-oxygen cycle, and it converts UV light into heat, preventing it from reaching the surface of the Earth.

Without the sun, the earth wouldn't have an ozone layer—but without the sun, the earth also wouldn't need it.[4]

The premise of greenhouse theory is that 'greenhouse gases' absorb and radiate infrared energy while regular air does not. This radiation is believed to provide a second source of heat for the earth, thus raising its temperature.

If, however, the earth is heated by the sun and also by the radiant transfer performed by trace gases, then its surface temperature must necessarily be higher than what sunlight can induce, due to the extra radiant energy impinging on it at any time of the day. Yet the simple fact is that the earth's sunlit surface temperature is entirely consistent with solar irradiance alone—which likewise means that greenhouse theory is demonstrably false on that point alone. The infrared radiated

---

[4] http://science.howstuffworks.com/earth5.htm

by trace gases cannot heat other air molecules, for they do not respond to infrared, nor is there any evidence of a 'second source' heating the earth.

Chapter 10

WHY?
by Alan Siddons

THE 19TH CENTURY saw the first clear articulation of radiative forcing theory…

> *The radiation of the sun in which the planet is incessantly plunged, penetrates the air, the earth, and the waters; its elements are divided, change direction in every way, and, penetrating the mass of the globe, would raise its temperature more and more, if the heat acquired were not exactly balanced by that which escapes in rays from all points of the surface and expands through the sky.*
> —Joseph Fourier (1768-1830)

The direct corollary of Fourier's conjecture is that less outgoing radiation will keep driving the temperature up. That's the essence of his theory, which has survived up to the current day. Indeed, Fourier regarded a glass enclosure as a real-life forcing model. Since glass is shortwave-transparent and infrared-opaque, he concluded that a garden greenhouse lets in visible light but prevents the 'dark rays' of infrared from escaping. Thus, he believed, the sun-induced heat inside a glass box was unable to escape, an imbalance which forced

81

the temperature to rise. Not so, it turns out, but Fourier's theory persisted even after this practical example was shown to be wrong.

The idea of trapping light was intriguing, however, and Gustav Kirchhoff (1824-1887) conceived a solution: a hole in a cave. A beam of light could enter this hole but the walls inside would absorb any reflections and prevent the light from escaping. Thus, by confining incoming radiation, the thermal energy which light confers could be shown to its maximum advantage. Kirchhoff's scheme was superior to selectively transmitting glass because a cave absorbs and traps all wavelengths of light, thus creating a complete radiative imbalance. At least theoretically.

Well, so what was found by cavity experiments? That a perfectly absorptive ('black') body rises to a temperature a bit higher than an actual black body that's open to convective heat loss from its surroundings. A theoretical blackbody thereby defines the upper limit of temperature vs. radiant absorption.

Try to grasp the implication, then. A blackbody cavity mimics the radiative restriction that 'greenhouse gases' are said to induce. Indeed, virtually none of the thermal radiation generated inside this cavity is allowed to escape. It 're-circulates' instead, and is sampled through a tiny hole. Does this confinement lead to a runaway greenhouse effect, though? No, it only sets an upper temperature limit—the SAME limit that's applied to the earth in the first place, for its estimated temperature is based on a blackbody equation!

Now, it is very likely that applying a cavity-based formula to the temperature of a rotating half-lit sphere is inherently mistaken. But if not, then 279 Kelvin constitutes the upper limit for the earth because such an estimate assumes a body that is perfectly absorptive, meaning that it can't possibly absorb more light than the light it's exposed to. Doing everything a 'greenhouse effect' is alleged to do; continuously re-radiating infrared energy inside itself, a light-trapping blackbody demonstrates that radiative forcing is a fiction. For its temperature hits a ceiling not much higher than what you see in real life. Yet greenhouse theory claims that radiative restriction generates temperatures HIGHER THAN a blackbody's. And considerably higher at that. Such a claim overtly contradicts experimental evidence, then. It doesn't have an empirical leg to stand on.

First seized upon as the answer and later dismissed, a glass enclosure proved that infrared opacity had nothing to do with generating extra heat inside. Then came the radiatively restricted blackbody, which nailed the forcing concept shut.

Yet, against all evidence, climatologists still push the radiative forcing theory.

Why?

Really new trails are rarely blazed in the great academies. The confining walls of conformist dogma are too dominating. To think originally, you must go forth into the wilderness.
—S. Warren Carey

Chapter 11

Analysis of Climate Alarmism
Part 2
by Tim Ball

Philosophical context

MOST PEOPLE HAVE no idea how their view of the world is unique to their culture and determined by values prevailing in society. It is inculcated through their upbringing including parental influence, religion and education. Like all philosophies that come to dominate society, climate hysteria is part of an evolution of ideas and needs an historical context.

Nowadays everything is about change as if it is something new. Western science and therefore basic education is developed from societies prevailing philosophy. Currently this assumes change is gradual over very long periods of time. Actually, rapid and significant change is the norm. This has allowed natural change and their rate to be identified as unnatural. Of course, there are always natural events so there are endless daily series of examples.

In Darwin's time, the English church accepted Bishop Ussher's biblically-based calculation that the world was formed on October 23, 4004 BC. But Darwin needed a much older world to allow the sort of evolution he envisioned as driving natural change. Simplistically, religion said God created the world in seven days; Darwin needed millions. Sir Charles Lyell provided the answer in a book titled, *Principles of Geology*, which Darwin took on his famous voyage to the Galapagos Islands. The combination of long time frames and slow development resulted in a philosophical view known as *uniformitarianism*.

If such a term sounds more appropriate to religion than science, that is because it is, in essence, another form of belief system. Uniformitarianism is the idea now underpinning western society's view of the world.

A basic tenet assumes change is gradual over long periods of time and any sudden or dramatic change is unnatural. Employing a version of uniformitarianism adapted to their needs, environmental extremists can point to practically any change and say it is unnatural, which implies it is man-made. But we know from modern science that natural changes can indeed be quite sudden and extreme—as Professor Tim Patterson of Carleton University, in Ottawa pointed out:

> *Ten thousand years ago, while the world was coming out of the thousand-year-long "Younger Dryas" cold episode, temperatures rose as much as 6 degrees C in a decade—100 times faster than the past century's 0.6 degrees C warming that has so upset environmentalists.*[1]

---

[1] http://www.financialpost.com/story.html?id=597d0677-2a05-47b4-b34f-b84068db11f4&k=29751

Happening as it did before the dawn of civilization, it was, of course, entirely natural.

Loss of credibility of science is serious at any time, but especially now when a major shift in philosophy is happening—what academics call a paradigm shift. We are moving from the end of the Scientific and Technological Revolution to a new order, or view of people and the planet.

The Scientific Revolution began in 1543 with a reluctant revolutionary Nicolas Copernicus, presenting a theory about the solar system. He replaced the earth (geocentric) at the center with the sun (heliocentric). This began a long process of undermining the Catholic belief in the structure of the solar system outlined 2,000 years earlier by Aristotle.

Copernicus triggered renewed research in astronomy and mathematics that is still going on today with the work of Stephen Hawking and others. They are linked through the centuries by famous men of science including, Johannes Kepler, Galileo Galilei, Isaac Newton, and Albert Einstein. Their ideas profoundly changed the scientific view of the universe and our solar system, but for most people they were of little consequence. A late 1990s survey in Europe found 17% of the population still believes the sun orbits the earth, not as Copernicus had it. As long as the sun rises and sets everyday it doesn't matter what science believes. The same is true of Newton's findings about gravity. As long as a person doesn't fly off into space, it's of little daily or even yearly consequence, but it is of consequence in a longer and larger context.

Charles Darwin was also a reluctant revolutionary but he found, like Copernicus, that once the cork was out of the bottle it couldn't be replaced. The church was upset in both cases, realizing like all dominant authorities how ideas were the ultimate danger. But Darwin's ideas had much wider and more profound consequences because they spoke directly to all people. Copernican ideas were too vast for medieval and most modern minds to grasp and therefore were less threatened. The church tried to bring their concerns to earth by arguing that his statement about an infinite universe left no space for heaven. It's difficult for us to grasp how important this was for medieval people. The threat of excommunication, denial of all church rites including access to heaven, makes little sense otherwise. But very few people knew about Copernicus or the church's concerns.

An effective, but rarely-used argument these days is *reductio ad absurdum*, or reducing to the absurd. The church said Darwin's view proposed humans were descended from apes—virtually, your grandparents were gorillas. Unlike all previous scientific theories, they realized Darwin's theory spoke directly and personally to everyone. Previously, science was remote from most people's lives, mysterious, obscure, of little consequence, now it was in every home, every church, and every school.

The scientific debate shifted away from the amoral, rational, logical to became variously religious, moral, philosophical or some combination, but always emotional.

Darwin's theory spawned a whole new school of study generally called the social sciences. Many believe this is—at best—a contradictory term, and at worst an oxymoron. The central theme of all the academic areas of sociology, political science, economics, psychology, anthropology, and human geography is the human animal. A specific segment included Social Darwinism, in which his scientific ideas of evolution, nature, and animals were applied to humans and human behavior.

In many ways these disciplines are contradictions because they try to show how humans are no different than the other animals, yet very different. The scientific view effectively rejected God as the reason for human existence on earth. Now, like all other animals, we were at the mercy of nature. We were no longer here for God's purpose so we didn't have His protection; we have to deal with nature and its threats on our own. Prior to formal religions, people's belief systems were collectively called animism and they revered natural objects such as the sun, moon, animals or birds. A deep-seated fear of nature and her ability to take lives underlies our concerns about environmental issues, especially global warming.

We emerged from the Cold War and the threat of nuclear annihilation with relief, although some believe the threat is still present. Many argue humans need an overwhelming presence of doom. If nature doesn't provide one, we create real or imaginary threats. Or is it as Raymond Aron said, "*In search of hope in an age of despair, the philosopher settles for an optimism based on catastrophe.*"

Threats of global warming or depletion of the ozone layer are more disturbing, because of their scale. There was always hope sense would intervene to avert a nuclear annihilation. Exploitation of these fears is compounded when governments say we can stop global warming, or repair the hole in the ozone. All we have to do is change our behavior and all will be well. This assumes we have accurate information about the problem, understand the mechanisms of the earth's systems, know the causes of the change and are capable of taking the correct remedial action without creating worse problems. With global warming, ozone depletion, and many other environmental issues, none of these conditions exist.

Science, with our compliance, has replaced God, leaving society to make the decisions and take actions to resolve problems. But even this is not the real issue. Religion is about morality, a code of living, which in most cases makes the individual or group accountable for their actions. Science is amoral, and essentially not accountable for its findings or actions. Society is left to deal with the moral and other questions that arise. Some scientists are aware of this dilemma and a few have warned society, usually without success. For example, Albert Einstein wrote to the President of the United States—warning of the potential dangers of nuclear power and urging politicians to show leadership in controlling the threat.

At the end of the 20th century people enjoyed the advances of science and technology, but negative side-effects were becoming apparent in some instances. In most cases, there were no scientific or technological solutions, the 'technological fix' was not an option. Now the issues required a moral answer, but these were thrust on a morally-confused society. Well, not everyone! Those with very fundamentalist religious views had no problem, often aggravating the issue by taking a 'holier than thou' position. Most realized they needed a moral position, but didn't want the one offered by the fundamentalist groups.

Some turned away from one organized religion to another—the green movement. Here was a nice, simple, morally-superior, non-religious solution. Stop your immoral behavior and all will be well. Return to the respectful ways of 'primitive' peoples from today and yesterday. The errors in this position require a book or two. The dilemmas and moral conflicts created for the green religion when 'primitive' people want the benefits of science and technology or resurrect traditional ways, such as whale hunting, are increasing every year. One daring challenge is found in Shepard Krech III's book *The Ecological Indian*.

So we have reached a midpoint in the transition from one paradigm to another. The religion of science replaced formal religion, but in doing so became more dogmatic than the religion it replaced. This is happening because there is a moral vacuum during the shift; a situation where political circumstances allow demagogues to advance their simplistic, undemocratic ideas that usually cause untold damage before sanity prevails.

Global warming is perhaps the extreme example of a victim of the current moral and intellectual vacuum. Most people incorrectly believe it is a change in climate due to human interference and confuse it with the Greenhouse Effect. They also believe both are new phenomena that are the result of impacts of the industrial world.

The Y2K fiasco was a fitting end to the 20th century. Predictions of doom and gloom following computer failure and subsequent technological collapse all proved to be totally incorrect. Despite a vigilant search by media around the world, no problems were found; the transition from one century to the next went without incident. Some governments claimed it was because of their vigilance, but this was simply an idle attempt to justify unwarranted expenditures.

The same governments warned that the greatest problems would occur in less developed nations such as Russia, China, and India because of antiquated computers. These countries spent virtually no money and had no problems, which proves the predictions were wrong and expenditures unnecessary.

This story is symptomatic of the 20th century that has been called the Age of Information, but is more properly called the Age of Misinformation, although the Age of Speculation is as good. During the 1990s, someone speculated that most computers, especially those running large public systems such as utilities, transport, and banking would not recognize the change from 1999 to 2000. This would cause them to shut down creating social, economic, and political chaos across the world. Books on the subject quickly appeared and media that thrive on threats of impending doom raised concern amongst the public to almost hysterical levels.

The exploiters who skillfully played on people's natural fears of impending disaster quickly silenced anyone raising a voice of reason. Concerns reached a level where politicians were forced to react. The squeaky wheel got the grease as usual, but only if it was environmentally friendly. They directed government departments to establish policies of remediation for the public and private sectors. In most cases, this involved the establishment of separate units to proof the system against any potential problem. This had three major effects:

• Nobody within government was determining if the problem was real;

• It gave the theory credibility because special interest groups argued that the government would not have established the units and provided funding if there wasn't a real problem;

• These units had a personal interest in perpetuating their jobs rather than saying there was no problem. Remember, it was a child who pointed out that the emperor had no clothes; the adults protected the self-interest of survival.

In this way a speculative theory developed into a prediction while avoiding rigorous intellectual and practical challenges. The truth came at 2359 hours on December 31, 1999 when all computer clocks around the world changed to the new millennium with no problems. The adage 'time will tell' was appropriate, specific, and finite—the doomsayers were completely wrong.

The Y2K problem has already slipped into oblivion—a fate that will befall most other 'predictions' of doom in the age of speculation.

I can hear the doomsayers shouting, "What if you're wrong?"

What they're really saying is "Shouldn't we act anyway?"

The answer is: not necessarily, but extremists use the blunt weapon of fear to cancel the use of calm, objective, reasonable options. The idea of acting 'just in case' is known as the Precautionary Principle and has merit in some instances. However, it assumes there is some clear relatively-uncontested evidence.

We cannot and should not act on every possible threat because it's not possible and it's not a 'no risk' world. We must work to reduce risk, but this requires prioritizing risks, and that requires some clear, relatively uncontested evidence. The fact is, science can speculate on a long list of potential dooms, but all that does is challenge society to decide which issues need attention. Using fear and creating hysteria makes it very difficult to make calm rational decisions about which issues need attention.

I used the following example to illustrate this point to the Parliamentary Standing Committee on the Environment regarding ozone. It was very clear the politicians did not understand that science works by presenting a hypothesis, which is then tested by other scientists.

My presentation began by listing some scientific facts:
• The earth was slowing in its speed of rotation;
• The magnetic field has weakened gradually and consistently over the last several decades. If this trend continues, the magnetic field will reach zero in approximately 120 years;
• When the earth's magnetic field disappears as it has done many times, mass extinction of species occurred.

I wanted to know what action my government planned for this impending disaster. Immediately one member expressed outrage at my presentation—pointing out that the issue was ozone. He completely missed my point and compounded his error by protesting how Galileo would be ashamed of me.

As a scientist, I was pursuing the deductive scientific method identified by Thomas Kuhn.[2] This

[2] Thomas S. Kuhn 1962, *The Structure of Scientific Revolutions*, University of Chicago Press.

means taking a collection of facts and attempting to develop a hypothesis linking and explaining them. I could have developed such hypotheses all day about a series of impending disasters, but this does not make them real or true.

In the other scientific method, a theory is developed and then tested in the laboratory all with facts gathered in the field. Kuhn called this the inductive method. It's rare for either method to exist in a pure form, but in both cases they are challenged and rigorously tested.

The theory is proved, proved with modifications, or rejected. If proved, at some point it will become a law of science, but this can take a long time. It requires that predictions made by the theory prove correct—the ability to predict is good definition of science.

Sir Isaac Newton, in his *Principia Mathematica*, included the theory of gravity, yet today we talk about the law of gravity. There was no conference at which scientists gathered to say it had been a theory long enough. The transition occurred when the theory made accurate predictions; and there is the key, because a very simple definition of science is the ability to predict. This raises interesting questions about weather forecasts, but more of that later.

Albert Einstein's theory of relativity was published in 1904 but remains a theory over one-hundred years later. Some predictions have proved correct yet science continues to have reservations and withholds the designation of law. Hesitancy speaks to another important part of the scientific method. Every hypothesis, whether inductive or deductive, is based upon a set of assumptions. They are both strength and weakness and become a point of attack in most cases.

The other goal is to gather facts that either support or destroy the hypothesis; or as T.H Huxley said:

The great tragedy of science—the slaying of a beautiful hypothesis by an ugly fact.

The most famous formula in science e $= mc^2$ is logically derived from Einstein's assumptions. The letters 'c' represents the speed of light and Einstein assumed nothing in the universe could travel faster than light.

In the year 2000, a scientific paper was published reporting the discovery of something traveling faster than the speed of light. If correct, the Einstein's theory is seriously weakened and the formula could become a footnote in scientific history.

Charles Darwin published his theory on the evolution of species in 1859. It remains a theory today for several reasons, but most importantly because it has never been seriously challenged by science. Darwin was, by default, chosen as the scientist whose work would finally overcome the power of religion.

Science began the conflict with the revolutionary ideas of Copernicus—and the struggle continued into the 20th century. Today we have the religion of science that has become more dogmatic than the religion it replaced. Any scientist who challenged Darwin would provide ammunition for the enemy. Creationists would leap on the opportunity to denounce evolutionary theory and replace it with creationism.

The creationist/evolution debate continues today in the work of Richard Dawkins...in his 1986 book *The Blind Watchmaker* and his 2006 book *The God Delusion*.

Scientists continue to create hypotheses using inductive and deductive methods, but now there is a disturbing development effectively preventing science from being science. The normal sequence of theory—followed by challenge and testing—has been short-circuited.

Very few journalists have any scientific training, but that wouldn't matter since they are seeking stories that fit the prevailing environmental hysteria of the day. Articles that seem to reinforce the global warming hypothesis usually receive attention while those contradicting or raising serious questions are avoided.

The media piece usually receives a high profile and is reinforced by information of little relevance except to skillfully-influence the public. For example, a story on the change in frequency of hurricanes will begin with reference to global warming when that subject isn't even mentioned in the original article.

Over the years I was always amazed by what stuck in the mind of the public about an issue. They invariably believed something was a proven fact or that a successful prediction was made. Most of the time there were no facts—only estimates—and no predictions, only theories.

What happened to cause the transition?

A vigilant, but unscientific monitoring of media stories on environmental issues seems to provide the answer. Most journalists include the conditional words and phrases necessary in the original scientific work; words such as *could* and phrases like *it appears that* usually appear in the story. The problem is they are taken in but not recorded by the public. What they remember is the headline in newspaper or a single statement at the beginning of the newscast. Invariably, these are simple, positive, unconditional statements—often changing the story from estimates to fact, and theory to prediction. If the story appears on television and in the newspaper, the repetition reinforces the accuracy and credibility of the story.

Special-interest groups take the information, usually without reference to the original article, and include it in their campaign—sometimes making it the sole focus of their propaganda.

Skilful manipulation exaggerates the potential threat, ignores scientific limitations and exploits people's fears so an objective search for the truth is no longer possible. Frequently, the level of concern leads to public demand for action—and politicians are left with little choice.

A steady campaign of propaganda in public meetings and rallies perpetuate and expand the fears. The issue is so widely discussed that most people are not willing to even entertain the idea that it is not true. Those who seized the moral high ground silence opponents. Government involvement that should serve to put the issue in perspective usually fuels the hysteria. National and international conferences occur with the democratic but illogical cast of characters ranging from the well-informed to the poorly-informed to the deliberately-misinformed.

Hysteria, emotionalism, and much hand wringing go on, but too often the wrong decisions are taken.

What are Weather and Climate?

Weather is the general atmospheric conditions experienced momentarily. Climate is the average of weather conditions for a region over a period of time.

As Robert Heinlein said:

Climate is what you expect, weather is what you get.

Climatology, the study of climate, is a generalist discipline and derives from the Greek word *klimat* referring to the angle of the sun or angle of incidence. It studies the average patterns of weather in a region or over time. Using the idea, the ancient Greeks determined there were three climate zones, Hot, Temperate and Frigid.

Climate is an anachronism in this age of specialization. It encompasses so many subjects, areas, and data, all ultimately interrelated—even the most sophisticated and powerful computers in the world are inadequate to incorporate even the most simplistic models.

Understanding this explains why so many people from diverse backgrounds and very specialized areas feel qualified to say they are climatologists. A simple analogy is climate as a jigsaw puzzle with thousands of pieces and each specialist with a single piece. Many claim their piece is essential to the entire puzzle. It also means others use a different piece but use it incorrectly or out of context. This book examines pieces of the puzzle used in attempts at climate reconstruction then specialists show how they were used incorrectly.

Continuing the puzzle analogy, but relating it to the process of solving a puzzle, puts the current level of climate science in context. We know the four corners of the climate puzzle are; the Sun, the Atmosphere, the Earth and the Oceans, but we are far from even minimal understanding of any one of them.

Next you locate the edge pieces and with climate this further underlines the limitations of current knowledge and research. The remaining pieces are separated into piles by color, but even here, some colors are definitive but many pieces have two and sometimes three colors, while others are shades and gradations. These are the pieces that invariably connect distinct areas and in climate it is the interaction between different segments that are important.

Another problem with a jigsaw puzzle is it is static while the real world is a constantly changing panorama. This is the great challenge for climate models beyond having enough pieces and a minimal understanding of the mechanisms of interaction and movement. For example, the computer models deal with the Earth as a flat disc bathed under a continuous twenty-four-hour haze of sunlight, ignoring the complexities created by a curved surface with an alternating day and night.

Of course, a puzzle or a model can have a cartoon quality. Cartooning is the art of providing minimal information to provide a general sense of the entire picture. How few lines do you need to recognize the individual? This doesn't work for climate and especially climate models, yet much is left out either because we don't know or there is insufficient computer capacity. I recall attending conferences in which the person who spoke loudest and claimed accurate results was the person with the biggest computer—the advent of the Cray computer was one such point.

However, there is another problem because you may have a piece of information but set it aside as insignificant. Yet, in a varying set of conditions the thresholds may be very different. This speaks to the problem of interaction and influence. For example, in the 1980s trace minerals (zinc, manganese etc.) in soil were considered of little consequence in the push to add or replace minerals used in crop production. It turned out that the plants ability to assimilate major chemicals required the catalyst of some trace minerals. When they were exhausted, plant yields declined. Climate change alarmists have exploited this concept of thresholds that they call tipping points. They ignore the problem when it comes to their computer models because inadequate computer capacity leaves them no choice.

In a strange historical twist, most people in the 20th century knew about meteorology before they knew about climatology. It's odd because meteorology is the study of physics of the atmosphere, a specific part or subset of climatology. Aristotle wrote a book titled *Meteorology* that was concerned with the processes and phenomena of the atmosphere. His intent was to understand mechanisms for weather forecasting.

This declined until the 19th century when development of instruments such as the thermometer and barometer combined with a desire to measure and understand the constituents of the atmosphere. An early example was discovery of oxygen independently by Scheele in 1773 and Priestley in 1774. Physics became more dominant so that by the beginning of the 20th century it dominated meteorology. In Canada, for example, to become a government weather forecaster, a Masters degree in Physics was required. After which a brief in-house course taught weather forecasting. There was virtually no climate instruction. The pattern was similar around the world.

Momentum came from attempts to measure and understand the atmosphere and the interactions that create weather. Meteorology's ascendancy continued during World War I as pilots needed accurate forecasts.

Tim Ball

It's why most weather stations are at airports and now suffer from interference from growing urban centers. Climatology gained attention in the academic world through the work of various people like Reid Bryson in the U.S., Kenneth Hare in Canada, Mikhail Budyko in Russia and Hubert Lamb in England. It only came to public attention when it became political in the late 1980s.

Climate came back on to the world stage at the height of dominance of specialization in the academic and research world. This development is critical to understanding the approach taken in this book. As specialists from outside climate science begin to look at the use of knowledge and understanding from their area, they realize the limitations, misapplication and errors created in the final conclusions about global warming and climate change.

The term renaissance person or polymath is someone who is skilled in many fields or disciplines with a breadth of knowledge and understanding. Benjamin Franklin is a good example. Some list Aristotle, but he falls into the category of universal person, that is, someone who knows all known science and geography. Alexander von Humboldt is generally considered the last universal person and he died in 1859. It's an auspicious year because it was the year Charles Darwin published his science- and world-changing work *On the Origin of Species*. Proliferation of science, triggered by Darwin's work, meant that nobody could encompass all scientific knowledge. Darwin was a natural philosopher, a term that preceded the modern designation of scientist. The change occurred with the introduction of the scientific method

102

that involved acquiring knowledge through experiments.

History of Weather and Climate Research

Early climate studies involved the work of geologists and glaciologists explaining the evidence of existence of massive ice sheets during the most recent Ice Age.

Louis Agassiz talked about the existence and extent of the ice sheets in Europe as early as 1837 but it was not generally accepted until the middle 1860s.[3] Early climate studies were attempts to explain the growth and retreat of the ice sheets. There were many theories, but one that endured was Joseph Adhemar's proposal that it was likely due to changes in the way the earth moves around the sun.[4] Through a series of additions and variations, Adhemar's idea evolved into the Milankovitch Effect, which is a combination of changes in the sun/earth relationships including varying orbit, tilt and date of equinox.

This is very important because this major mechanism of climate change is not included in current Intergovernmental on Climate Change (IPCC) computer model calculations used as the basis of world energy, environmental and economic policy.

Hubert Lamb is generally considered the father of modern climatology. During World War II he realized the inadequacy of weather forecasts and he used time on

---

[3] Imbrie, J, and Imbrie, K. P., *Ice Ages: Solving the Mystery*, New Jersey (page 46), Enslow Publishers, 1979.

[4] Adhemar, J. A., 1842, *Revolutions de la mer: Déluges Périodiques,*, Paris

the night shift searching the archives of the United Kingdom Meteorological Office (UKMO) for greater insight. His theory was that a better understanding of past weather patterns would allow for better forecasts. He soon discovered the extent to which climate varies even within historic times.

After the war, he established the Climatic Research Unit (CRU) at the University of East Anglia with the goal of gathering information on past weather from the vast variety of direct and secondary sources called proxy data. In 1977 he produced the comprehensive classic two volume set, *Climate: Past, Present and Future.*[5] I was privileged to have Lamb help with my doctoral thesis and act as reviewer on an early article. He would be mortified by what has happened at the CRU, but not surprised. Lamb knew what was going on because he cryptically writes in his autobiography, *Through all the Changing Scenes of Life: A Meteorologists Tale,* how a grant from the Rockefeller Foundation came to grief because of:

> *...an understandable difference of scientific judgment between me and the scientist, Dr. Tom Wigley, whom we have appointed to take charge of the research.*

---

[5] Lamb, H.H., *Climate: Past, Present and Future*, London, Methuen and Co, LTD, 1977.

Figure 1: Wigley and H.H. Lamb, Founder of the CRU[6]

Wigley became the Director of the CRU prior to moving to a position in the U.S. Phil Jones replaced him as Director and was in charge through the period covered by the now infamous leaked emails that disclosed the manipulation and corruption that makes this book necessary. It's obvious from the emails that Wigley is the grandfather figure controlling the corruption of climate science…his career is a classic example of what is wrong with climate science. Educated as a mathematical physicist he gravitated to climate and carbon-cycle modeling. His National Center for Atmospheric Research, Boulder, USA biography says:

His main interests are in carbon cycle modeling, projections of future climate and sea-level change and

---

[6] http://www.cgd.ucar.edu/cas/symposium/

*interpretation of past climate change particularly
with a view to detecting anthropogenic influences.*

His training has nothing to do with any of those topics
and many of the problems in climate science are related
to misuse of computer models. Most troubling is his
focus on anthropogenic influences because that has
apparently colored his science.

Political Control of Climate Research

How and why did climate shift from a scientific study
into a political issue? All this provides context for
specific examples of corrupted science used to underpin
the political science.

Weather was always a factor in short-term planning
and sometimes very significant in the directions of
history. For example, with two thousand years in
between and going in different directions, Julius
Caesar's invasion of England in 55 A.D. and the Allied
invasion of Europe from England in 1944 were both
hampered by bad weather. Climate change is the major
control over the Earth's history and therefore human
history. Primary influence is by varying production of
food supply. This is important because it raises
questions about concern over temperature and warming
when precipitation is a much more important variable.

Climate and specifically temperature became a
political consideration as recently as the 1970s.
Ironically, the concern was global cooling because from
1940 to 1980 global temperatures declined. This
cooling period was to become a problem for the later
hysteria over global warming. The threat of cooling saw

a shift back into conditions experienced during the Little Ice Age (LIA). This cold period was from approximately 1450 to 1850 with a nadir in the 1680s.

Media sensationalists and alarmists saw an opportunity, as they have done throughout the climate debate of the last thirty years. Alarmist books such as Lowell Ponte's *The Cooling*[7] was classic. Consider this from the preface:

> *It is cold fact: the global cooling presents humankind with the most important social, political, and adaptive challenge we have had to deal with for ten thousand years. Your stake in the decisions we make concerning it is of ultimate importance; the survival of ourselves, our children, our species.*

It begins with a false premise and then appeals to emotions by threatening the children. It's a pattern repeated many times since. A team of journalists produced *The Weather Conspiracy*[8] subtitled *The Coming of the New Ice Age*. On the cover it asks an appropriate question but then essentially ignores it in the book:

> *Have our weather patterns run amok? Or are they part of a natural and alarming timetable?*

They seek credibility by adding a gold sticker advising that the book includes two CIA Reports.

---

[7] Lowell Ponte, *The Cooling*, New Jersey, Prentice-Hall, 1976.
[8] Impact Team Report, *The Weather Conspiracy*, New York, Ballantine Books, 1977.

Themes developed by the CIA reports are representative of the academic and political thinking of the day. One report titled, *Potential Implications of Trends in World Population, Food Production, and Climate*[9] argues:

> *Trying to provide adequate world food supplies will become a problem of over-riding priority in the years and decades immediately ahead. ... Even in the most favorable circumstances predictable, with increased devotion of scarce resources and technical expertise, the outcome will be doubtful; in the event of adverse changes in climate, the outcome can only be grave.*

In the *Climate* section is the comment essential to the analysis in this chapter:

> *Far more disturbing is the thesis that the weather we call normal is, in fact highly abnormal and unusually felicitous in terms of supporting agricultural output. While still unable to explain how or why climate changes, or to predict the extent and duration of change, a number of climatologists are in agreement that the northern hemisphere, at least, is growing cooler.*[10]

Overpopulation, inadequate resources, especially food supply are central to claims of the Club of Rome

---

[9] Directorate of Intelligence, Office of Political Research, OPR-401, August 1974.
[10] Op cit.

formed in 1968 and set out in the 1972 report *The Limits to Growth.*[11]

Their 1974 report titled, *Mankind at the Turning Point*[12] says:

> *It would seem that humans need a common motivation...either a real one or else one invented for the purpose.... In searching for a new enemy to unite us, we came up with the idea that pollution, the threat of global warming, water shortages, famine and the like would fit the bill. All these dangers are caused by human intervention, and it is only through changed attitudes and behavior that they can be overcome. The real enemy then, is humanity itself.*

Several themes developed from the Club of Rome became the center of social and political trends since 1960. Understanding how climate science was distorted and perverted since then can only be considered in that context. Chief among these was the new paradigm of environmentalism. It is not a coincidence that Paul Ehrlich, leading scientist involved with the Club and author of the book *The Population Bomb*, was creator of Earth Day. A major part of the challenge people faced who questioned the hypothesis that humans were causing global warming was the charge that they didn't care about the environment, the children and the

---

[11] Meadows, D. H., Meadows, D. I., Randers, J., Behrens, W.W., *The Limits to Growth, A Report to the Club of Rome*, 1972.

[12] Mihajlo Mesarovic and Eduard Pestel, *Mankind at the Turning Point, A Report to the Club of Rome*, E.P. Dutton, 1975

future. Occupation of the moral high ground inhibited any who dared to question. This worked against the fundamental function of scientists to be skeptics. In a complete reversal of normality those who did were labeled skeptics demonstrating how little most understand science and the scientific method.

A simple definition of science is the ability to predict. Every prediction they made has been completely wrong. Most incorrect are the population predictions followed by rate of reduction of available resources. This is very important to the climate issue because the IPCC temperature scenarios for the future are based on the assumption that population predictions are accurate and the rate of consumption of fossil fuels will increase in parallel.

Population, or at least overpopulation, is central to arguments of environmentalism and the Club says that we have to stop growth, especially growth engendered by fossil fuels. We need a new world order. One member of the Club with the contacts and organizational abilities to tackle such an incredible objective was Maurice Strong. In 1990 he said:

> What if a small group of these world leaders were to conclude the principal risk to the earth comes from the actions of the rich countries?...In order to save the planet, the group decides: Isn't the only hope for the planet that the industrialized civilizations collapse? Isn't it our responsibility to bring this about?

He told *Maclean's* magazine in 1976 that he was "*a socialist in ideology, a capitalist in methodology.*" Presumably this justifies the duplicity in a socialist making a great deal of money as an industrialist. He also warned that, *"...if we don't heed his environmentalist warnings, the Earth will collapse into chaos."*

The challenge is converting the idea to a reality. How do you shut down industrialized nations? An analogy helps understand how Strong and a few like-minded people did it. Compare the nation to a car and think about how you can stop the engine. You can squeeze the fuel line and starve the engine, however, if you did that in any country people would react quickly and negatively. However, you can stop an engine by plugging the exhaust.

Strong's method is not a physical stop—as you do with an engine, but a metaphorical stop. If you can show that one part of the industrial exhaust is causing catastrophic global warming putting the survival of the planet in jeopardy you have your instrument. It's even better if you can use science to make the case.

You need two components to carry out your plan. One is a scientific agency; the other is a global political agency that can bypass national governments. Strong's experience told him the United Nations (UN) was his vehicle. Elaine Dewar, wrote about Strong in her book *"Cloak of Green"*[13] and concluded that he liked the UN because:

He could raise his own money from whomever he liked, appoint anyone he wanted, control the agenda.

---

[13] Elaine Dewar, *Cloak of Green: The Links between Key Environmental Groups, Government and Big Business*, James Lorimer & Company, LTD, Toronto.

Tim Ball

The challenge was twofold. Advance the political agenda and provide the scientific evidence to provide legitimacy. Organization of and appointment as first Secretary General of the United Nations Environmental Program established in 1972 provided the political platform. Out of that agency and in conjunction with the World Meteorological Organization (WMO) the Intergovernmental Panel on Climate Change (IPCC) was formed to provide and advance the scientific evidence.

As they note on their web site[14]:

> The Intergovernmental Panel on Climate Change (IPCC) was established by WMO and the United Nations Environment Programme to assess scientific, technical and socio-economic information relevant for the understanding of climate change, its potential impacts and options for adaptation and mitigation.

This is the group touted as the consensus on climate change research. It is anything but, and has been a political agency from its inception, but it has convinced the public that humans, especially their CO_2, are causing climate change by continuing to publish periodic reports.

Other events were providing the fertile social and political ground needed to further the goals. Anything that would suggest human activities and particularly industry were causing environmental problems became a focus. A report Strong commissioned for the first UNEP conference and prepared by Barbara Ward and

---

[14] http://www.wmo.int/pages/partners/ipcc/index_en.html

Rene Dubos titled, *Only One Earth: The Care and Maintenance of a Small Planet* essentially became the first state of the environment report.[15]

It contained political catch phrases that became the *lingua franca* of environmentalism such as Dubos' *"Think globally, act locally"* or the Brundlandt Commission's *"Sustainable development."*

The latter is a typical political statement because it means everything to everyone and nothing to anyone.

Environmental and special interest groups received a world platform and ascendancy by receiving Consultative Status at the 1992 conference Strong organized and chaired in Rio de Janeiro. The idea of Consultative Status was resurrected along with the concept of Non-Governmental Organizations (NGO) from original ideas incorporated in the UN Charter.

The conference was dubbed the Earth Summit, but as with the current debate, large segments of society, including industry and business, were essentially excluded. They were subsequently given token status by establishment of the little-known World Business Council for Sustainable Development (WBSCD). One critical piece of the objective was established at the Conference to further Strong's agenda of controlling climate science through politics; the Climate Change Convention out of which the Kyoto Accord emerged.

Now everything was in place to control the science and further the political agenda. Now policies could evolve, but because they were based on incorrect science would have devastating consequences. Now the

---

[15] Barbara Ward and René Dubos, *Only One Earth: The Care and Maintenance of a Small Planet*, W.W. Norton Company, New York, 1972

challenge was to perpetuate the misinformation and divert scientists, who, despite personal attacks, denial of funding, and exclusion from national and world level conferences, continued to pursue the scientific method.

WMO involvement in research about weather and climate is logical, but it's hard to understand why they are doing political research. The IPCC Reports were not the first because they did research and produced reports on the global cooling concerns of the 1970s. Martin Parry was one person involved in those reports and in the formation of the IPCC.

A photograph (Figure 3) shows him in Villach Austria in 1985, when the formation of the IPCC was given substance.

It is an important photograph because it shows Tom Wigley and Phil Jones, who were already involved with the Climatic Research Unit (CRU), connected to the IPCC they controlled. The WMO connection was essential because it meant all governments were directly involved and thereby controlled through their weather agencies. It put all the power in the hands of the bureaucrats who could control their government's policies. Politicians were loath to question bureaucratic scientists who purportedly knew of what they were saying.

They could also wait until a politician was replaced. All those involved, though ostensibly bureaucrats, behaved with greater political guile and deception than any politicians, but with none of the accountability.

Strong's powerful connections in Canada included personal friendship and obligations from Canadian Prime Minister Paul Martin.[16] It is not surprising that

---

[16] http://www.canadafreepress.com/2004/cover120904.htm

Canadian, Gordon McBean, who later became Assistant Deputy Minister (ADM) of Environment Canada (EC), chaired the 1985 Villach, Austria meeting.

Figure 3: Gathering of people involved with formation IPCC[17]

The most influential bureaucrat appointed to the IPCC was Sir John Houghton. His career as Chief Executive of the United Kingdom Meteorological Office (UKMO) overlapped with his role as first co-chair of the Intergovernmental Panel on Climate Change (IPCC) and lead editor of the first three Reports. His political bias was evident throughout his tenure. He denies saying, *"Unless we announce disasters no one will listen."* He claims he would have said, *"There are those who will say*

---

[17] http://www.cgd.ucar.edu/cas/symposium/

'unless we announce disasters, no one will listen', but I'm not one of them[18]

If that's the case how does he explain the article he wrote titled, *Global warming is now a weapon of mass destruction,* which includes the claim that *it kills more people than terrorism.*[19]

This is scientifically incorrect and grossly irresponsible. More people die of cold each year than warm. Houghton's position typifies the emotional, political position typical of those associated with the IPCC. He was appointed as a scientist but was clearly chosen because of his political bias.

His co-chair Bert Bolin was scientifically qualified as a Professor of Meteorology but had a history of involvement in environmental politics. He and Houghton signed the 1992 warning to humanity essentially blaming the developed nations. It was more of the Club of Rome approach with no clear measures or evidence, simply a list of possible disasters if we didn't do things their way.

Consensus was a major argument in support of the claim that humans were the primary cause of global warming—almost from the inception of the IPCC. It was coincident because it was the people involved with the IPCC that was the consensus. Appointment of people who would support the goals of the IPCC was essential and Houghton and Bolin were two classic

---

[18] http://www.independent.co.uk/environment/climate-change/fabricated-quote-used-to-discredit-climatescientist-1894552.html
[19]

http://www.guardian.co.uk/politics/2003/jul/28/environment.greenpolitics

candidates. This is wholly contrary to established scientific principles that require all scientists to question and challenge the hypothesis under discussion. Despite this, the IPCC claims it conducts a purely scientific investigation and implies all involved are climate experts. Nothing could be further from the truth.

There are almost always extreme dissenters, but general views are usually well supported by the science. This is not true with climate science. Claims of consensus on the climate change issue are not valid or applicable in science. Consensus is not a scientific fact, however, it is important in politics and this underlines the political nature of the climate change issue and the role of the IPCC.

A headline from the UN reads:

> Evidence is now 'unequivocal' that humans are causing global warming — UN report.[20]

They are talking about the Fourth Assessment Report (FAR) of the Intergovernmental Panel on Climate Change (IPCC), but unfortunately they begin with false information. In a subtle exploitation of the consensus argument they incorrectly write:

> The IPCC, which brings together the world's leading climate scientists and experts.

John McLean disabuses this argument.

---

[20]

http://un.org/apps/news/story.asp?newsID=21429&cr=climate&cr1=change

The IPCC would have us believe that its reports are diligently reviewed by many hundreds of scientists and that these reviewers endorse the contents of the report. An analysis of the reviewers' comments for the scientific assessment report by Working Group I show a very different and very worrying story.[21]

Or as MIT Professor Richard Lindzen, former member of the IPCC said:

It is no small matter that routine weather service functionaries from New Zealand to Tanzania are referred to as 'the world's leading climate scientists.' It should come as no surprise that they will be determinedly supportive of the process.[22]

Madhav Khandekar a former employee of Environment Canada and expert 2007 IPCC reviewer in a letter to the Ottawa Hill Times wrote:

Brant Boucher, in his letter "Scientific consensus" (The Hill Times, Aug. 6, 2007), seems to naively believe that the climate change science espoused in the Intergovernmental Panel on Climate Change IPCC documents represents "scientific consensus." Nothing could be further from the truth! As one of the invited expert reviewers for the 2007 IPCC

[21]

http://findarticles.com/p/articles/mi_m0EIN/is_2007_Sept_10/ai_n19506379

[22]

http://www.heartland.org/policybot/results/1069/IPCC_report_criticized_by_one_of_its_authors.html *documents, I have pointed out the flawed review process used by the IPCC scientists in one of my letters (The Hill Times, May 28, 2007).[23]*

Participants and structure of the IPCC were honed to establish a specific outcome. As Richard Lindzen, Professor of Meteorology at MIT, said, they were supportive of the process. Now it was necessary to predetermine the outcome. A favorite political technique to preserve the appearance of openness yet retain control is to allow a commission of inquiry. You then control the inquiry by limiting the investigation through definitions and terms of reference.

Science works by creation of theories based on assumptions, in which scientists performing their proper role as skeptics, try to disprove the theory. The structure and mandate of the IPCC was in direct contradiction to this scientific method. They set out to prove the theory rather than disprove it. The AGW theory was proposed and almost immediately accepted as fact.

All efforts focused on proving instead of trying to disprove the theory. As Karl Popper explains:

"One can sum up all this by saying that the criterion of the scientific status of a theory is its falsifiability, or refutability, or testability." He also notes that, "It is easy to obtain confirmations,

---

23

http://thehilltimes.ca/page/view/.2007.august.13.letter4

or verifications, for nearly every theory — if we look for confirmations.[24]

Most people have no idea what the IPCC actually studies. They believe their reports are complete reports of climate change. This misconception is mostly because the IPCC arranged it that way. In fact, they only look at that portion of climate change caused by humans. This is how they limit their study.

Note that the United Nations Framework Convention on Climate Change (UNFCCC), in its Article 1, defines climate change as:

a change of climate which is attributed directly or indirectly to human activity that alters the composition of the global atmosphere and which is in addition to natural climate variability observed over comparable time periods.[25]

The UNFCCC thus makes a distinction between climate change attributable to human activities altering the atmospheric composition, and climate variability attributable to natural causes. This makes the human impact the primary purpose of the research. The problem is you cannot determine the human portion of climate change if you don't know how much it changes naturally—and we don't. The IPCC assumes humans

---

24

http://www.stephenjaygould.org/ctrl/popper_falsification.html

[25] http://www.ipcc.ch/pdf/assessment-report/ar4/syr/ar4_syr_appendix.pdf

cause most of the changes that are occurring and set out to prove that is true.

Properly, a scientific definition would put natural climate variability first, but at no point does the UN mandate require an advance of climate science. The definition used by UNFCCC predetermined how the research and results would be political and predetermined. It made discovering a clear 'human signal' mandatory, but essentially meaningless.

Other parts of their mandate illustrate the political nature of the entire exercise. Its own principles require the IPCC:

> ...shall concentrate its activities on the tasks allotted to it by the relevant WMO Executive Council and UNEP Governing Council resolutions and decisions as well as on actions in support of the UN Framework Convention on Climate Change.[26]

The role is also to:

> ...assess on a comprehensive, objective, open and transparent basis the scientific, technical and socio-economic information relevant to understanding the scientific basis of risk of human-induced climate change, its potential impacts and options for adaptation and mitigation. IPCC reports should be neutral with respect to policy...

---

[26] Principles Governing IPCC work, approved at the 14th Session, Vienna October 1-3, 1998 and amended at the 21st Session, Vienna November 6-7, 2003.

The cynicism of this last sentence is that they then made the Summary for Policymakers (SPM) the most important part of IPCC reports—and these summaries are anything but neutral.

The IPCC is a political organization and is the sole basis of the claim of a scientific consensus on climate change.

Consensus is neither a scientific fact nor important in science, but it is very important in politics. There are 2,500 members in the IPCC divided between 600 in Working Group I (WGI)—the actual climate science. In the most recent report in 2007 only 308 of the 600 worked on the science part of the report and only five reviewed all eleven chapters. The remaining 1,900 in working Groups II and III (WG II and III) all study impacts. They accept without question the findings of WGI and assume warming due to humans is a certainty.

In a circular argument typical of so much climate politics, the work of the 1,900 is listed as 'proof' of human caused global warming. Through this they established the IPCC as the only credible authority and consensus thus further isolating those who raised questions.

The manipulation and politics didn't stop there. The Technical Reports of the three Working Groups are set aside and another group prepares the Summary for Policymakers (SPM). A few of the scientists prepare a first draft, which is then reviewed by government representatives. These scientists effectively control the SPM and always included key people all later identified in the email scandal known as Climategate. A second draft is produced, and then a final report is hammered out as a compromise between the scientists and the individual government representatives. It was this process that allowed Michael Mann, author of the infamous and scientifically-flawed 'hockey stick', to be a lead author of the Paleoclimate (Chapter 2)[27] section and of the SPM for the 2001 IPCC Report.

It's not surprising that the 'hockey stick' is front and centre in the SPM. It became pivotal evidence in convincing politicians and the public that the present was warmer than the past. Section 2.3 is titled, *Is the recent warming unusual?* This question confronts the challenge made by the few skeptics with a public voice about past climate. Control of the research done by the IPCC was supplemented by the need to counteract growing critiques of their claims.

The SPM is then released at least three months before the science report. Most of the scientists involved in the science report see the Summary for the first time when it is released to the public. The time between its release to the public and the release of the Technical Report is taken up with making sure it aligns with what the politicians/scientists have concluded. Here is the instruction in the IPCC procedures:

> *Changes (other than grammatical or minor editorial changes) made after acceptance by the Working Group or the Panel shall be those necessary to ensure consistency with the Summary for Policymakers (SPM) or the Overview Chapter.*

This is like an Executive writing a summary and then having employees write a report that agrees with the summary.

---

[27] http://www.grida.no/publications/other/ipcc_tar/

When you accept a hypothesis before it is proven, you step on the treadmill of maintaining the hypothesis. This leads to selective and even biased research and publications. As evidence appears to show problems with the hypothesis, the natural tendency is to become more virulent in defending the increasingly indefensible.

This tendency is underlined by John Maynard Keynes sardonic question:

If the facts change, I'll change my opinion. What do you do, Sir?

The IPCC and those who were chosen to participate were locked in to a conclusion by the rules, regulations and procedures carefully crafted by Maurice Strong. These predetermined the outcome—a situation in complete contradiction to the objectives and methods of science.

As evidence grew that the hypothesis was scientifically unsupportable, adherents began defending the increasingly indefensible rather than accept and adjust. The trail they made is marked by the search for a clear human signal, identified in modern parlance as 'smoking guns'.

They turned increasingly to rewriting history and producing biased results—thus expanding the gap between what they claimed and what the evidence showed. As explained above, this was done even within the structure by the gap between what the Working Group I: The Scientific Basis, Report was reporting and the political message of the SPM.

It started it early. The main report is then reviewed to make sure it 'aligns' with the summary. Here again is the instruction in the IPCC procedures.

> *Changes (other than grammatical or minor editorial changes) made after acceptance by the Working Group or the Panel shall be those necessary to ensure consistency with the Summary for Policymakers (SPM) or the Overview Chapter.*

Of course, even minor editorial changes can be problematic. In 1995, Chapter 8 lead author Benjamin Santer made such changes to accommodate the SPM to the political—in contradiction to the agreed text. Why would you appoint scientific experts to write separate portions of a technical report then have them 'adjust' their information or views to fit a summary?

The most logical illogical conclusion is the SPM is the political portion of the document and the scientific experts are expected to conform. Maybe the simple answer is it is not a summary. We now know Santer was a very important part of the Climategate scandal. He was a graduate of CRU and his 1983 thesis, supervised by Tom Wigley, used Monte Carlo methods in the validation of climate models.

The Chapter 8 controversy involved the most important part of all IPCC reports, namely, the evidence for a 'human signal'.

It was a search Santer was directed to by Professor Klaus Hasselmann during his post-graduate employment at the Max-Planck Institute. As Wigley's protégé he was the perfect candidate for the IPCC.[28]

Chapter 8 didn't have specific evidence of a human signal. The original draft submitted by Santer read:

> *Finally we have come to the most difficult question of all: "When will the detection and unambiguous attribution of human-induced climate change occur?" In the light of the very large signal and noise uncertainties discussed in the Chapter, it is not surprising that the best answer to this question is, "We do not know."*

This was changed by Santer to accommodate the SPM to read:

> *The body of statistical evidence in Chapter 8, when examined in the context of our physical understanding of the climate system, now points toward a discernible human influence on global climate.*[29]

Notice this is 'statistical evidence', not actual evidence, but is part of the growing desire to 'blame' humans.

Compare it with the comment in the 1990 IPCC report:

> *...it is not possible at this time to attribute all, or even a large part, of the observed global-mean*

---

[28]

http://www.sepp.org/Archive/controv/ipcccont/Item03.htm

[29] http://icecap.us/images/uploads/Ben_Santer.pdf

warming to (an) enhanced greenhouse effect on the basis of the observational data currently available.

By the time of the 2001 report, the politics and hysteria had risen to a level that demanded a clear signal. A large number of academic, political, and bureaucratic careers had evolved and depended on expansion of the evidence.

Meanwhile, personal attacks and isolation of skeptics was in full swing. Clear evidence was provided in the Technical Report by a tree ring study published in 1998 by Mann, Bradley and Hughes, (known as MBH98). Mann was a lead author on the SPM and the graph, descriptively named the 'hockey stick,' was prominently displayed. This raised serious concerns about the objectivity of a Summary with major input from scientists citing their own research. Unfortunately, this is typical of the incestuous political nature of the entire IPCC process.

The hockey stick fiasco was unmasked by a basic scientific test known as reproducible results. Other scientists use the same data and procedures to replicate the original findings. McIntyre and McKitrick (M&M) attempted, but failed to reproduce, the MBH98 findings. A debate ensued with claims M&M were wrong and unqualified climate experts. They replied that Mann had refused to disclose all the codes he used to achieve the results, but even without them the major problem was a misuse of data and statistical techniques.

An important point to make at this juncture relative the theme of this book is that McIntyre knew nothing about climate and wasn't even interested. He was at conference in which the hockey stick graph was shown. From his experience with statistics and plotting graphs he knew immediately how the data and methods were misused.

The U.S. National Academy of Sciences (NAS) appointed a committee chaired by Professor Wegman to investigate and arbitrate. His committee report found in favor of M&M:

> *It is not clear that Mann and associates realized the error in their methodology at the time of publication. Because of the lack of full documentation of their data and computer code, we have not been able to reproduce their research. We did, however, successfully recapture similar results to those of MM. This recreation supports the critique of the MBH98 methods, as the offset of the mean value creates an artificially large deviation from the desired mean value of zero.[30]*

Most people, especially in the media, missed the equally startling and disturbing conclusion by Wegman.

> *In our further exploration of the social network of authorships in temperature reconstruction, we found that at least 43 authors have direct ties to Dr. Mann by virtue of co-authored papers with him. Our findings from this analysis suggest that authors in the area of paleoclimate studies are closely connected and thus 'independent studies' may not be as independent as they might appear on the surface.*

---

[30]

http://republicans.energycommerce.house.gov/108/home/07142006_Wegman_Report.pdf

Wegman's Report preceded disclosure of the activities at the Climatic Research Unit and how they peer-reviewed each other's work and controlled peer review by intimidating editors even to the point of having one fired for publishing an article they didn't like.

But what was the objective of the hockey stick research?

There were hundreds of research papers from a wide variety of sources confirming the existence of a period warmer than today known as the Medieval Warm Period (MWP). This period was clearly warmer than present temperatures and warmer than some computer model predictions. Its existence was a serious problem because it negated the claims that the 20th century temperatures were unprecedented.

What to do?

Even before the emails were leaked we had one part of the answer and that was to rewrite history. Professor Deming wrote the following letter to *Science*:

> *With the publication of the article in Science [in 1995], I gained significant credibility in the community of scientists working on climate change. They thought I was one of them, someone who would pervert science in the service of social and political causes. So one of them let his guard down. A major person working in the area of climate change and global warming sent me an astonishing email that said "**We have to get rid of the Medieval Warm Period**. (Emphasis added)*

The hockey stick graph showed no temperature increase for 1,000 years (the handle) with a sudden upturn in the 20th century (the blade). Besides misusing data and statistical methods it also overrode a vast array of research from a variety of sources that established the existence of the Medieval Warm Period.

So, the hockey stick was scientifically inaccurate, but it served to remove threats to the anthropogenic global warming theory.

The second action was revealed after the emails were leaked—an orchestrated attack on Soon and Baliunas, authors of an excellent work confirming the existence of the Medieval Warm Period (MWP) from a multitude of sources.[31]

Their work challenged attempts to get rid of the MWP because it contradicted the claim by the proponents of Anthropogenic Global Warming (AGW). Several scientists challenged the claim that the latter part of the 20th century was the warmest ever. They knew the claim was false; many warmer periods occurred in the past. Michael Mann 'got rid' of the MWP with his production of the hockey stick, but Soon and Baliunas were problematic. What better than have a powerful academic destroy their credibility for you? Sadly, there are always people who will do the dirty work.

A perfect person and opportunity appeared on October 16, 2003. Michael Mann, infamous for his lead in the 'hockey stick' that dominated the 2001 IPCC Report, sent an email to people involved in the CRU scandal:

---

[31] Soon, W. and Baliunas, S., 2003, Proxy climatic and environmental changes of the last 1000 years. Climate Research, 23, 89-110.

Dear All, Thought you would be interested in this exchange, which John Holdren of Harvard has been kind enough to pass along...

At the time Holdren was the Teresa and John Heinz Professor of Environmental Policy & Director, Program in Science, Technology, & Public Policy, Belfer Center for Science and International Affairs, John F. Kennedy School of Government.

He is now Director of the White House Office of Science and Technology Policy, Assistant to the President for Science and Technology and co-Chair of the President's Council of Advisors on Science and Technology—informally known as the United States *Science Czar.*

In an email on October 16, 2003 from John Holdren to Michael Mann and Tom Wigley we are told:

I'm forwarding for your entertainment an exchange that followed from my being quoted in the Harvard Crimson to the effect that you and your colleagues are right and my "Harvard" colleagues Soon and Baliunas are wrong about what the evidence shows concerning surface temperatures over the past millennium. The cover note to faculty and postdocs in a regular Wednesday breakfast discussion group on environmental science and public policy in Harvard's Department of Earth and Planetary Sciences is more or less self-explanatory.

This is what Holdren sent to the Wednesday breakfast group:

I append here an e-mail correspondence I have engaged in over the past few days trying to educate a Soon/Baliunas supporter who originally wrote to me asking how I could think that Soon and Baliunas are wrong and Mann et al. are right (a view attributed to me, correctly, in the Harvard Crimson). This individual apparently runs a web site on which he had been touting the Soon/Baliunas position.

The exchange Holdren refers to is a challenge by Nick Schulz editor of Tech Central Station (TCS). On August 9, 2003 Schulz wrote:

In a recent Crimson story on the work of Soon and Baliunas, who have written for my website [1] www.techcentralstation.com, you are quoted as saying: My impression is that the critics are right. It' s unfortunate that so much attention is paid to a flawed analysis, but that's what happens when something happens to support the political climate in Washington. Do you feel the same way about the work of Mann et al.? If not why not?

Holdren provides lengthy responses on October 13, 14, and 16, but the comments fail to answer Schulz's questions. After the first response Schulz replies:

I guess my problem concerns what lawyers call the burden of proof. The burden weighs heavily much more heavily, given the claims on Mann et al. than it does on Soon/Baliunas. Would you agree?

Of course, Holdren doesn't agree. He replies:

> *But, in practice, burden of proof is an evolving thing—it evolves as the amount of evidence relevant to a particular proposition grows.*

No it doesn't evolve; it is either on one side or the other. This argument is in line with what has happened with AGW.

He then demonstrates his lack of understanding of science and climate science by opting for Mann and his hockey stick over Soon and Baliunas. His entire defense and position devolves to a political position. His attempt to belittle Soon and Baliunas in front of colleagues is a measure of the man's blindness and political opportunism that pervades everything he says or does.

Schulz provides a solid summary when he writes:

> *I'll close by saying I'm willing to admit that, as someone lacking a PhD, I could be punching above my weight. But I will ask you a different but related question. How much hope is there for reaching reasonable public policy decisions that affect the lives of millions if the science upon which those decisions must be made is said to be by definition beyond the reach of those people?*

We now know it was deliberately placed beyond the reach of the people by the group that he used to ridicule Soon and Baliunas. He was blinded by his political views, which as his record shows are frightening. One web site synthesizes his position on over-population as follows, *Forced abortions. Mass sterilization. A "Planetary*

Regime" with the power of life and death over American citizens.[32]

The hockey stick elimination of the MWP solved one problem for the AGW proponents, but 'scientific' support for the blade was required.

It was provided in the same 2001 IPCC report by the Director of the CRU, Phil Jones, with the claim of an increase of 0.6°C in the global average annual temperature in 130 years. They claimed the increase is beyond any natural increase and is therefore anthropogenic.

This is simply incorrect.

The figure was promoted by the SPM and the media, but what it actually said was the increase was 0.6°C ±0.2°C, an error factor of 66%. This puts it well within the error factor of global average temperatures estimates.

In addition, there are so many problems with the data that as McKitrick shows, it is impossible to calculate a global annual temperature.[33][34]

Here are some of these problems:

• There are very few records of 130 years length;
• There are fewer stations now than in 1960;
• Most of these are concentrated in eastern North America and Western Europe;

---

[32] http://zombietime.com/john_holdren/
[33] http://rossmckitrick.weebly.com/uploads/4/8/0/8/4808045/surfacetempreview.pdf
[34] McKitrick, R., Essex, C., Andresen, B., *Does a Global Temperature Exist*, Journal of Non-Equilibrium Thermodynamics, Volume 32, No. 1

- Most of these stations are affected by the Urban Heat Island effect;
- There are virtually no measurements for the oceans that are 70% of the surface.

There is serious scientific concern about the nature, length and quality of the database best expressed by the U.S. National Research Council Report in 1999:

> *Deficiencies in the accuracy, quality and continuity of the records place serious limitations on the confidence that can be placed in the research results.*

Kevin Trenberth, member of the IPCC and leading member of the CRU group commented:

> *It's very clear we do not have a climate observing system ... This may be a shock to many people who assume that we do know adequately what's going on with climate, but we don't.*

On October 14, 2009, Trenberth wrote in one of the leaked emails that exposed climate science corruption he said:

> *The fact is that we can't account for the lack of warming at the moment and it is a travesty that we can't. The CERES data published in the August BAMS 09 supplement on 2008 shows there should be even more warming: but the data are surely wrong. Our observing system is inadequate.*

These remarks are troubling for Jones, but they are even more problematic for constructing global climate models.

But there was a more serious problem with Jones' results because he refused to disclose which stations he used and how the data was adjusted.

To a request from Warwick Hughes, an Australian researcher who long sought to verify the global temperature record Jones wrote:

> We have 25 or so years invested in the work. Why should I make the data available to you, when your aim is to try and find something wrong with it." (Jones' reply to Warwick Hughes, 21. February 2005; P. Jones later confirmed this.)

More problematic is the fact we will never know because Jones admits the data is now lost.[35] [36]

Apparently Jones is not alone in the practice of non-disclosure or denial of access to climate data. A series of failed attempts to obtain information from the University of East Anglia and from the joint enterprise of the Hadley Centre and the CRU known as HadCrut3 are well documented on Steve McIntyre's ClimateAudit website.[37]

---

[35]

http://www.guardian.co.uk/environment/2010/feb/15/phil-jones-lost-weather-data

[36] http://www.cato.org/pub_display.php?pub_id=10578

[37] http://www.climateaudit.org

The Data is Critical Yet Woefully Inadequate

In a previous section, the work of Thomas Kuhn was mentioned. It speaks to the structure of scientific revolutions and identifies the two basic approaches. The inductive method has a scientist create a theory and then seek data to prove or disprove it. The deductive method is used when the scientist has data and then works out an explanatory theory. With either method, the amount and accuracy of the data is critical.

A major criticism of the hockey stick is that it blended data from two different sources. This speaks to the ongoing problem of climate research, namely the type and quality of data available.

Generally there are three different areas of climate reconstruction that approximate a time scale and each yields data of different accuracy and reliability. The first, and most recent, is the instrumental or secular period that covers approximately one-hundred years. It's assumed this provides the most accurate record, though global annual average temperatures and other temperature data are provided at least two decimal places, they are statistics derived from instrumental readings—all taken to a half-degree accuracy at best.

The second source of data is the historic record that covers the period from which human observations are available: approximately 3,000 years. Most of this data is derived from proxy or secondary sources such as dates of harvest or first snowfall. Temperature approximations within one degree Celsius are the best one can expect.

The third source is biologic and geologic evidence—which covers the vast amount of the earth's history. Apart from the degree of temperature accuracy, which is greater than one degree Celsius, there is the problem of accurate dating. Climate reconstruction requires accurate juxtaposition of data. Even the most sophisticated technique, radiocarbon dating, only covers approximately 70,000 years with an error factor that increases as you go back in time. The most common technique for the geologic record is potassium/argon (K/A). Here the error factor is a major problem. For example:

> Potassium-argon dates usually have comparatively large plus or minus factors—they may be on the order of .25 million years for a 2 million year old date.[38]

To put this in perspective, just consider how much climate has changed in the last 250,000 years—it covers a complete cycle from interglacial to full glacial and back to interglacial again.

Climate proxy indicators that transcend two areas are valuable and tree rings (dendroclimatology) is one. It was the main technique used for the hockey stick, but was grossly misused. For example, they assumed that growth depicted in the rings is purely a measure of temperature. In fact, for most trees, precipitation is a much more important factor. Second they overlapped the tree ring reconstruction with the modern temperature record and that is unrealistic. It compares with the overlaps attempted between ice core records

---

[38] http://anthro.palomar.edu/time/time_5.htm

and the modern atmospheric CO_2 readings. Ernst Georg Beck identified this problem when he compared ice core data with the modern Mauna Loa measures and 19th atmospheric measures (Figure 4).

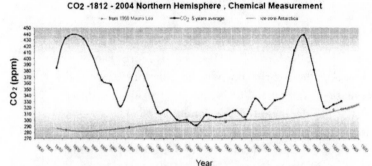

CO2 -1812 - 2004 Northern Hemisphere , Chemical Measurement

Figure 4: Comparison 138 yearly averages of CO_2 with ice core and Mauna Loa.[39]

No wonder the IPCC were driven to confess that:

> *Current spatial coverage, temporal resolution and age control of available Holocene proxy data limit the ability to determine if there were multi-decadal periods of global warmth comparable to the last half of the 20th century.*[40]

That the public was led to believe climate change is new speaks to the issue of historic climates. All the evidence, crude as it is, indicates that change currently occurring

---

[39] http://www.anenglishmanscastle.com/180_years_accurate _Co2_Chemical_Methods.pdf
[40]

http://www.ipcc.ch/publications_and_data/ar4/wg1/en/ ch6s6-5-1-3.html is well within natural variability. Most mistakenly think the modern record is better. It isn't—it is also a victim of the manipulation and distortions done to prove that humans are the cause of climate change.

The Instrumental Record

There are serious concerns about data quality in the instrumental record. The U.S. spends more than other areas of the world on weather stations, yet their condition and reliability is simply atrocious. Anthony Watts has documented the condition of U.S. weather stations; it is one of government's failures.[41]

Figure 5 shows quality ratings for stations in the US Historical Climate Network (USHCN). 69% of stations (CRN=4 and =5) have error ranges equal to or greater than 2°C. Only 10% (CRN=1 and =2) have errors less than 1°C.

Figure 5: Weather Station Quality Rating.[42]

---

[41] http://www.surfacestations.org/
[42] Op cit.

Evidence of manipulation and misrepresentation of data is everywhere. Countries maintain weather stations and adjust the data before it's submitted through the World Meteorological Organization (WMO) to the central agencies including the Global Historical Climatology Network (GHCN)[43], the Hadley Center associated with CRU now called CRUTEM3[44], and NASA's Goddard Institute for Space Studies (GISS)[45]. They make further adjustments before selecting stations to produce their global annual average temperature.

In a valuable paper titled, *A Critical Review of Global Surface Temperature Data*, McKitrick provides a very good analysis that underscores the problems[46].

> The number of weather stations providing data to GHCN plunged in 1990 and again in 2005. The sample size has fallen by over 75% from its peak in the early 1970s, and is now smaller than at any time since 1919. The collapse in sample size has not been spatially uniform. It has increased the relative fraction of data coming from airports to about 50 percent (up from about 30 percent in the 1970s). It has also reduced the average latitude of source data and removed relatively more high-altitude monitoring sites. GHCN applies adjustments to try and correct for sampling discontinuities. These have tended to increase the

[43] http://www.ncdc.noaa.gov/ghcnm/
[44] http://www.cru.uea.ac.uk/cru/data/temperature/
[45] http://data.giss.nasa.gov/gistemp/
[46] http://climateresearchnews.com/2010/08/a-critical-review-of-global-surface-temperature-data-products-by-ross-mckitrick/

> *warming trend over the 20th century. After 1990*
> *the magnitude of the adjustments (positive and*
> *negative) gets implausibly large.*

This is why they produce different measures each year from supposedly similar data.

James Hansen controls the records maintained and adjusted by NASA GISS. He was the scientist who put the entire issue of warming in the public and political arena when he appeared before Al Gore's Senate Committee in 1988.

Remember what his former boss John Theon said. It was very pointed and apparently an implication about what Hansen was doing…

> *Furthermore, some scientists have manipulated the*
> *observed data to justify their model results. In*
> *doing so, they neither explain what they have*
> *modified in the observations, nor explain how they*
> *did it.*

They all use data derived from the GHCN and are consistently different from those of other agencies. Under Hansen's control GISS 'adjustments' and errors always produce higher temperatures. They limited eligible stations (Figure 5). Only approximately 1,000 stations have 100 records. A dramatic decrease in the number of stations after 1960 and diminished global coverage affected the global temperatures. McKitrick has shown how the reduction in numbers of stations creates another artificial temperature change after 1990. (Figure 6).

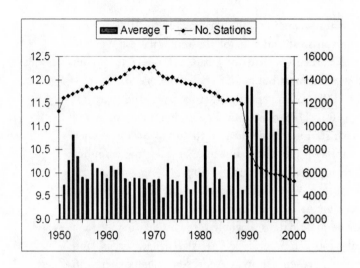

Figure 6: Global weather stations versus simple mean of the temperature.[47]

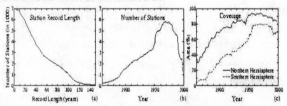

Figure 7: GISS graphs showing a) station record length b) # of Stations c) Global coverage.

Examples of GISS errors illustrates why the numbers they produce are of no value. In 2007 a 'Y2K' error made 1998 the warmest year on record and nine of the

[47] http://www.uoguelph.ca/~rmckitri/research/nvst.html

ten warmest years in the U.S. record in the1990s. Now 1934 is warmest and four of the ten warmest were in the 1930s. Hansen said they had not made the claim 1998 was warmest but a GISS staffer disagreed.[48]

In 2008, GISS reported October was the warmest since 1880. They'd re-used September data for many northern stations. GISS blamed the error on the agency that supplied the data, but NASA said the supplier carried out 'extensive quality control'. As always, the error output was in the news, while the correction received virtually no mainstream attention[49].

Inadequate data is a problem for calculating average annual global temperature, as McKitrick, Essex and Bjarne Andresen show. Andresen, an expert in thermodynamics, explains:

> It is impossible to talk about a single temperature for something as complicated as the climate of Earth...a temperature can be defined only for a homogeneous system. Furthermore, the climate is not governed by a single temperature. Rather, differences of temperatures drive the processes and create the storms, sea currents, thunder, etc. which make up the climate.

So far this discussion has dealt with the inadequacies of the temperature data. Like the entire global warming and climate change diversion it was a singular focus on

---

[48] http://wattsupwiththat.com/2010/01/14/foiad-emails-from-hansen-and-giss-staffers-show-disagreement-over-1998-1934-u-s-temperature-ranking/
[49] http://climateaudit.org/2008/11/12/gavin-schmidt-the-processing-algorithm-worked-fine/

warming. The trouble is: weather involves a multitude of variables all of which are essential to explanation and understanding. These include, among others, wind direction and speed, barometric pressure, cloud cover, but especially moisture content of the air and precipitation.

Global temperature data is grossly inadequate but precipitation measures are even worse. They are inadequate in the modern record and virtually non-existent and impossible to recreate from the historic record. Consider this comment about Africa:

> One obvious problem is a lack of data. Africa's network of 1152 weather watch stations, which provide real-time data and supply international climate archives, is just one-eighth the minimum density recommended by the World Meteorological Organization (WMO). Furthermore, the stations that do exist often fail to report.[50]

The quote is from an article about trying to predict the critical monsoon in the Sahel region of Africa.

> Climate scientists cannot say what has delayed the monsoon this year or whether the delay is part of a larger trend. Nor do they fully understand the mechanisms that govern rainfall over the Sahel. Most frustrating, perhaps, is that their prognostic tools—computer simulations of future climate— disagree on what lies ahead. "The issue of where Sahel climate is going is contentious," says

---

[50] *Waiting for the Monsoon*, Science, August 2006, Volume 313.

Alessandra Giannini, a climate scientist at Columbia University. Some models predict a wetter future; others, a drier one. "They cannot all be right."

And that speaks to the bigger problem with the inadequate data because it is used to construct the computer models.

Data collection is expensive and requires continuity—it's a major role for government. They failed with weather data primarily because money went to political climate research. A positive outcome of corrupted climate science exposed by Climategate is re-examination beginning with raw data by the UK Met Office (UKMO)[51]. This is impossible because much is lost—thrown out after modification or conveniently lost; as in the case of records held by Phil Jones, head (Director of Research) of the University of East Anglia Climatic Research Unit.

Climate Models Were a Pivotal Change

The dramatic change in climate research came with the introduction of computer models. They appeared to provide the ability to deal with large volumes of data and the complex interactions between various components of the climate system. Instead they became the vehicle manipulated to produce the output necessary to support the political objectives.

I watched this trend as more and more, climate modelers dominated climate meetings and conferences.

---

[51] http://wattsupwiththat.com/2010/2/23/met-office-pushes-a-surface-temperature-data-do-over/

A remarkable occurrence at a 1987 conference in Edmonton, Alberta provides an example of this trend, but also an example of what is wrong with climate models and thereby climate science.

Keynote speaker Michael Schlesinger's address was titled *Model Projections of the Equilibrium and Transient Climatic Changes Induced by Increased Atmospheric CO_2*[52].

During his presentation there was much agitation from someone sitting behind me. After the presentation, many people asked rather angry questions—including a senior bureaucrat who asked about the accuracy of the predictions. Schlesinger replied about 50 percent to which the bureaucrat replied we are planning on planting trees in areas your projections show extreme aridity. My minister wants 98 percent.

After more raucous debate a shoe flew on to the stage from behind. In the silence that followed the agitated man behind me who had a voice box and could not get attention, said I did not have a towel. He then went to the stage announced his qualifications as an atmospheric physicist and wrote a formula on the blackboard. He asked if this was the basic formula used in the model to represent the atmosphere. When assured it was he began to eliminate variables each of which Schlesinger agreed was eliminated in the computer models. At the end he said what is left is meaningless as representative of the atmosphere.

---

[52] Magil, B. L., and Geddes, F., (editors) 1988. *The Impact of Climate Variability and Change on the Canadian Prairies: Symposium Workshop Proceedings*. Prepared by Alberta Department of the Environment, 1987 September 9-11, Edmonton, Alberta. 412p.

Schlesinger had taken the same data and run it through five different climate models.

> *Five recent simulations of CO_2-induced climatic change by atmospheric GCM/mixed-layer ocean models are contrasted in terms of their surface air temperature and soil moisture changes. These comparisons reveal qualitative similarities and quantitative differences.*

When asked what he meant by "*qualitative similarities and quantitative differences,*" he said the similarities were that all models showed increasing temperature but the amounts varied regionally. This speaks to the ongoing problem with climate models and climate science. Of course, they all show temperature increase because they are programmed to have temperature rise if CO_2 increases. The quantitative difference refers to the variability in temperature change from region to region. The problem is these differences are massive with one showing an entire continent different than another.

The climate models truly reflect the old acronym GIGO, Garbage In, Garbage Out. But they also benefit from the modern adulation and awe associated with computers and computer output. As Pierre Gallois explained:

> *If you put tomfoolery into a computer, nothing comes out of it but tomfoolery. But this tomfoolery, having passed through a very expensive machine, is somehow ennobled and no-one dares criticize it.*

In climate science, driven by the lack of data (discussed later), they used the artificial output of one computer model as real data in another model.

Some basic problems associated with IPCC use of computer models must preface any discussion about their use to persuade the public and the politicians that the output has validity.

• Models run by forcing variables under a variety of preset conditions. In climate models this has involved doubling CO_2;

• Each model run takes weeks of computer time, even though the computer is making millions of calculations a second. The run is complete when a new equilibrium is reached;

• Each time the same computer is forced with exactly the same conditions and starting at the same point a different result is reached;

• The final value used in reports, such as those of the IPCC, is the average of a series of runs;

• The IPCC use the average output of several different models;

• The IPCC does not make predictions. They call them 'scenarios';

• The scenarios are only partly based on physical processes—they assume future economic and social conditions; what Richard Lindzen described as:

> ...very much a children's exercise of what might possibly happen, prepared by 'a peculiar group' in

the IPCC almost all of whom have 'no technical competence.[53]

Initially, models were focused on weather and later on climate forecasting. The earliest efforts were simple numerical models until a breakthrough in the 1970s came with the work of Syukuro Manabe.

But, even then people warned about the limitations. Bert Bolin provided an early warning about the limitations when he wrote:

There is very little hope for the possibility of deducing a theory for the general circulation of the atmosphere from the complete hydrodynamic and thermodynamic equations[54].

This statement is still essentially true and part of the debate about climate-based strategy. In 1977 Abelson wrote about more apparent limitations:

Meteorologists still hold out global modeling as the best hope for achieving climate prediction. However, optimism has been replaced by a sober realization that the problem is enormously complex[55].

---

[53]

http://www.melaniephillips.com/diary/archives/000482.html

[54] Bert Bolin, *Studies of the General Circulation of the Atmosphere*, Academic Press, Advances in Geophysics 1, 87-116, 1952.

[55] Abelson, P. H., *Energy and Climate*, Science, September 2, 1977.

Spencer Weart explains a long sequence of attempts to improve the models—all without much success[56]. As leading climate modeler at the European Centre for Medium-Range Forecasts in Reading, England Tim Palmer said:

> *I don't want to undermine the IPCC, but the forecasts, especially for regional climate change, are immensely uncertain.*

He prefaced this with the comment:

> *"Politicians seem to think the science is a done deal.[57]*

The idea the science is settled was the claim of Vice President Al Gore and undoubtedly was galvanized in 1988 by the appearance before the joint House and Senate committee of James Hansen Director of NASA Goddard Institute of Space Studies (GISS).

Hansen said:

> *"the greenhouse effect has been detected and it is changing our climate now" and there was "a strong cause and effect relationship between the current climate and human alteration of the atmosphere."*

However, the most unsupportable claim was that we are:

---

[56] http://www.aip.org/history/climate/GCM.htm
[57]

http://www.newscientist.com/article/mg19826543.700-poor-forecasting-undermines-climate-debate.html

> *... 99 percent certain that the warming trend was not a natural variation but was caused by a build-up of carbon dioxide and other artificial gases in the atmosphere.* [58]

Later Hansen claimed he was muzzled, but his former boss, Dr. John Theon, head of NASA's Weather and Climate Research Program from 1982 to 1994, disavowed the lie.

> *Hansen was never muzzled even though he violated NASA's official agency position on climate forecasting (i.e., we did not know enough to forecast climate change or mankind's effect on it). Hansen thus embarrassed NASA by coming out with his claims of global warming in 1988 in his testimony before Congress.* [59]

Theon was even more pointed.

> *My own belief concerning anthropogenic climate change is that the models do not realistically simulate the climate system because there are many very important sub-grid scale processes that the models either replicate poorly or completely omit. Furthermore, some scientists have manipulated the*

---

[58] http://www.nytimes.com/1988/06/24/us/global-warming-has-begun-expert-tells-senate.html
[59]

http://epw.senate.gov/public/index.cfm?FuseAction=Minority.Blogs&ContentRecord_id=1A5E6E32-802A-23AD-40ED-ECD53CD3D320

*observed data to justify their model results. In
doing so, they neither explain what they have
modified in the observations, nor explain how they
did it.*

He later said Hansen's testimony was an embarrassment
to NASA because the official NASA position was that
they didn't understand the climate system well enough
to make a reliable forecast.

*"I don't have much faith in the models," Theon
says, pointing to the "huge uncertainty in the role
clouds play." Theon describes Hansen as a "nice,
likeable fellow," but worries "he's been overcome by
his belief—almost religious—that he's going to
save the world.*[60]

After Hansen's appearance, the issue of global warming
became increasingly political and the science more
directed to proving the theory that human production
of CO_2 was the cause. Government money was the
primary source of research funding and it served to
skew the research to proving—rather than
disproving—the theory. It is more than coincidence
that the Intergovernmental Panel on Climate Change
(IPCC) was formed in Villach, Austria in 1988. It was
all part of increasing political control of climate science.
Now the official source used by all governments of
climate change data and forecasts is the IPCC.

Weather is a very complex system. When you
stand outside in the weather what you experience is
what scientists call white noise. It is comprised of a

---

[60] Op cit.

multitude of red noises that include everything from cosmic radiation in deep space to volcanic heat from the bottom of the ocean and everything in between. Figure 8 is a simple diagram of the weather system showing major components and some of the interactions between components.

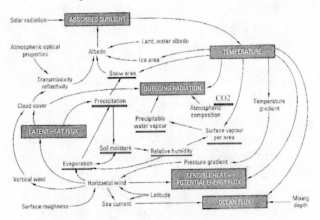

Figure 8: A simple diagram illustrating complexity of weather and climate.[61]

CO_2 is identified separately as part of the "Atmospheric Composition" category. It is one miniscule part. Water vapor is much more important part because it has direct impact on those items underlined in red.

Combine the systems diagram in Figure 8 with the inadequate data and the problems of climate science and lack of confidence in models is clearly justified. Figure 9 is a schematic showing the construct of a General Circulation Model (GCM).

[61] After Briggs, David J., Smithson, Peter, Ball, Timothy *Fundamentals of Physical Geography*, Toronto, Copp Clark Pitman (1989)

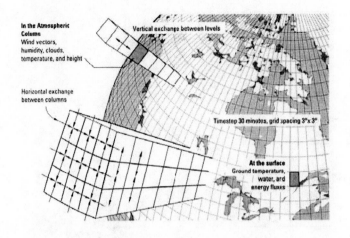

In the Atmospheric Column
Wind vectors, humidity, clouds, temperature, and height

Vertical exchange between levels

Horizontal exchange between columns

Timestep 30 minutes, grid spacing 3°x 3°

At the surface
Ground temperature, water, and energy fluxes

Figure 9: Schematic of GCM Showing Grid Structure.[62]

The model is actually a mathematical construct that represents the weather in each rectangle, but for most of the world, including the 70 percent that is ocean, there is no data. Figure 10 shows the distribution of weather stations in the GHCN file and the vast gaps that exist in the surface stations.

---

[62] Op cit.

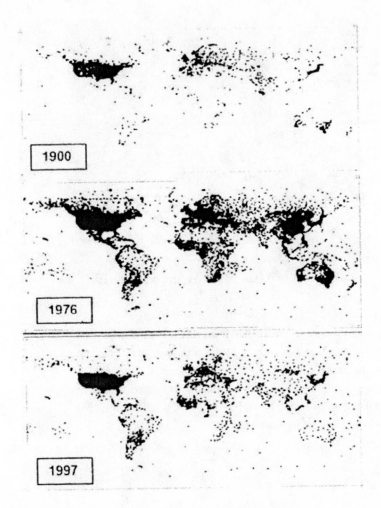

**Figure 10: Surface Weather Station Change
Comparison.**[63]

[63]http://scienceandpublicpolicy.org/images/stories/papers
/originals/surface_temp.pdf

An excellent study and analysis of the surface temperature record was performed by Joseph D'Aleo and Anthony Watts. Their summary is presented as in the original paper. It is an indictment of the record and the people responsible for its maintenance and analysis.

1. Instrumental temperature data for the pre-satellite era (1850-1980) have been so widely, systematically, and unidirectionally tampered with that it cannot be credibly asserted there has been any significant 'global warming' in the 20th century.

2. All terrestrial surface-temperature databases exhibit very serious problems that render them useless for determining accurate long-term temperature trends.

3. All of the problems have skewed the data so as greatly to overstate observed warming both regionally and globally.

4. Global terrestrial temperature data are gravely compromised because more than three-quarters of the 6,000 stations that once existed are no longer reporting.

5. There has been a severe bias towards removing higher-altitude, higher-latitude, and rural stations, leading to a further serious overstatement of warming.

6. Contamination by urbanization, changes in land use, improper siting, and inadequately-calibrated instrument upgrades further overstates warming.

7. Numerous peer-reviewed papers in recent years have shown the overstatement of observed longer term warming is 30-50% from heat-island contamination alone.

8.   Cherry-picking of observing sites combined with interpolation to vacant data grids may make heat-island bias greater than 50% of 20th-century warming.

9.   In the oceans, data are missing and uncertainties are substantial. Comprehensive coverage has only been available since 2003, and shows no warming.

10.   Satellite temperature monitoring has provided an alternative to terrestrial stations in compiling the global lower-troposphere temperature record. Their findings are increasingly diverging from the station-based constructions in a manner consistent with evidence of a warm bias in the surface temperature record.

11.   NOAA and NASA, along with CRU, were the driving forces behind the systematic hyping of 20th-century 'global warming'.

12.   Changes have been made to alter the historical record to mask cyclical changes that could be readily explained by natural factors like multidecadal ocean and solar changes.

13.   Global terrestrial data bases are seriously flawed and can no longer be trusted to assess climate trends or validate model forecasts.

14.   An inclusive external assessment is essential of the surface temperature record of CRU, GISS and NCDC "chaired and paneled by mutually agreed to climate scientists who do not have a vested interest in the outcome of the evaluations."

15.   Reliance on the global data by both the UNIPCC and the US GCRP/CCSP also requires a full investigation and audit.[64]

64

http://scienceandpublicpolicy.org/images/stories/papers/originals/surface_temp.pdf

But what do they do if there are no weather stations in one box of the grid? They simply go up to 1200km away to get data to apply to the entire area of the box. This is a huge problem in vast areas of Canada and Russia, which are critical to weather systems in the Northern Hemisphere.

Figure 11 shows the reduction in the number of stations in Canada.

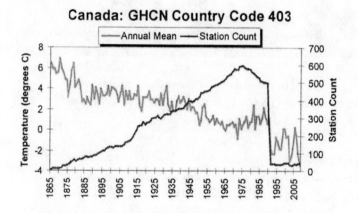

Figure 11: Dramatic Decline in Weather Stations in Canada.[65]

What is more dramatic is the number of Canadian stations currently used to calculate global average annual temperature as shown in Figure 12. Only those stations identified with a black diamond are used. Notice that only one station, Eureka, is used for the

---

[65]http://scienceandpublicpolicy.org/images/stories/papers/originals/surface_temp.pdf

entire northern half of Canada including the Arctic, Eureka is a station identified as a refugia because of local conditions and unique plant species.

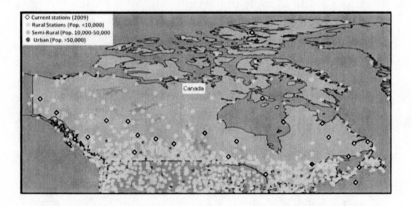

Figure 12: The Number of Canadian Stations Currently used to Calculate the Global Average Annual Temperature.

This illustrates the inadequacy of the surface record as the basis for the determining global average annual temperature, but also as the basis for a computer model.

The surface data is totally inadequate even without the manipulation, but there is a bigger problem. The atmosphere and therefore the model are three-dimensional and there is virtually no modern or historic data above the surface.

Conclusion

In book *True Enough: Learning to live in a Post-Fact Society*, Farhad Manjoo claims:

> *Facts no longer matter. We simply decide how we want to see the world and then go out and find experts and evidence to back our beliefs*[66].
> —Farhad Manjoo

The Post-Fact society is very active in environmental and climate issues. The inadequate, manipulated science of climate and climate change make it useless and unacceptable as the basis for any environmental, energy or economic policy. The media, which generally abandoned the concepts of fairness and balance, amplifies the manipulation. There's a constant search for examples to buttress the idea that global warming and climate change are caused by humans. The IPCC and supporters of their charade spend their time finding confirmations. In doing so, they defeated the scientific method by trying to prove the AGW theory, but as Karl Popper notes, *It is easy to obtain confirmations, or verifications, for nearly every theory—if we look for confirmations*[67]. Even after their arguments are proven inaccurate, inappropriate, or wrong, their ideas are pursued and new 'facts' are found.

A central mission is to find a human cause, and then exploit fear by claiming a failure to stop the activity will

---

[66] *True Enough: Learning to Live in a Post-Fact Society*, Farhad Majoo, John Wiley and Sons, New York, 2008.
[67] *Conjectures and Refutations: The Growth of Scientific Knowledge,* Karl Popper, Routledge Classics, New York, 2002.

result in planetary destruction. Governments exacerbate the problem by accepting the singular position that global warming is occurring due to human addition of carbon dioxide to the atmosphere. They now talk about climate change, but climate is always changing. They again moved the goalposts recently when John Holdren announced that we were experiencing *climate disruptions*[68]. Climate is always changing, the current evidence is now more supportive of a cooling trend.

So, the challenge is to prepare for what is more likely when the prevailing political wisdom asserts the opposite. Sir Walter Scott's observation, *Oh, what a tangled web we weave when first we practice to deceive*[69] is sadly appropriate.

The problem, as always, is the innocent people, not the guilty, wind up paying the price.

---

[68] http://motls.blogspot.com/2010/09/global-climate-disruption-holdren.html
[69] *Marmion, a Tale of Flodden Field*, Sir Walter Scott, 1808.

Chapter 12

A History of Encounters with the Sky Dragon
by Martin Hertzberg

The First Skirmish—a Blow Against Prudery

MY FIRST ENCOUNTER with the Sky Dragon occurred in the French Alps, but the first blow in that encounter was not mine but my wife's! It was at a NATO-sponsored meeting on coal combustion held in 1986 at Les Arcs.

My wife and I and three colleagues from MIT—and their very proper wives—were congregated at the swimming pool of the hotel where the meeting was being held. We were chatting about this and that when another colleague from Australia arrived to join us. Shortly thereafter, his girlfriend appeared *aux seins nus*; that is, bare-breasted in a topless bathing suit. She proceeded to dive into the pool and swim. We men pretended not to notice how well endowed she was as she swam backstroke before us, but the proper Bostonian wives were shocked.

Chatting among themselves, they proceeded to roundly condemn the young Australian lady for her scandalous behavior.

163

My wife and I listened to all the chatter. I sat quietly without saying a word—not daring to suggest it didn't bother me at all. My MIT colleagues did likewise, but my wife had heard enough. She proceeded to the ladies room and reappeared shortly, herself in a topless condition, and joined the young Australian lady in the pool both swimming bare-breasted.

Two things happened that evening at dinner.

First, my Australian colleague got up (you know how unpretentious those Australians are) and proposed a toast to my wife for her exceptionally well endowed swimming performance at the pool earlier that day.

Secondly, one of my MIT colleagues who had witnessed it all was so impressed that he solicited my opinion on the subject of greenhouse warming of the atmosphere by human CO_2 emission. He was on an NAS committee considering the question and had read a paper of mine presented at the Combustion Symposium at MIT. I had used the infrared emission from the 4.2 micron band of CO_2 to measure methane explosion temperatures in a twelve-foot-diameter sphere. He also apparently knew that I had once served a Meteorologist while on active duty with the U. S. Navy. Being asked for an opinion by someone from MIT is a great honor.

I responded that, although CO_2 was an essential ingredient for the photosynthesis that supports almost all life on Earth, I doubted such a minor constituent of the atmosphere could have a significant effect on the radiative balance between the Sun and the Earth. I also suggested that the overall role of the atmospheric 'greenhouse effect' could be checked by comparing the Earth's average surface temperature with that of the Moon. It receives essentially the same input radiance from the Sun but has no atmosphere.

Scouting the Enemy

In 1989, at a Symposium at Chatham College in Pittsburgh (formerly the Pennsylvania College for Women, Rachel Carson's alma mater), a paper was presented describing a model in which greenhouse gas induced temperature changes in the atmosphere were driving the Earth's ocean circulation. I had to heckle the speaker with the obvious fact that he had it 'backasswards'.

Meteorologists know from the El Nino phenomenon, the moderate temperatures in Western Europe caused by the Gulf Stream, the development and motions of hurricanes and typhoons, and the periodic summer monsoons in Asia and elsewhere, that it is the other way around; namely, the distribution of land and ocean and ocean currents drive the atmospheric circulation.

Clearly the model being presented had the 'tail wagging the dog'. In the same symposium, I had a brief discussion with a distinguished atmospheric scientist who during his presentation had repeated the standard mantra that the atmosphere of Venus was hot because of a 'greenhouse effect' caused by its high CO_2 content. When I asked him whether he had corrected for the adiabatic compression caused by its high surface pressure, he responded that that was only a small correction factor. I left the Symposium in disbelief: something was terribly wrong.

A short time later, I had a similar discussion with the then President of the Combustion Institute, who repeated that same mantra about the temperature of Venus. He informed me that he was on an NAS panel considering the global warming issue. When I asked him whether he had considered the effect of Venus' closer distance to the Sun, and the effect of adiabatic compression in its very dense atmosphere, I got a rather blank stare.

While he was a rather distinguished chemist, the conversation convinced me that I was better qualified than he was to be on that panel. After all, temperatures in regions below sea level such as Death Valley and the Dead Sea are higher than in surrounding areas at sea level because of adiabatic compression, and of course, those higher temperatures have absolutely nothing to do with the CO_2 content of our atmosphere.

Defeating the Sky Dragon

Shortly thereafter, a colleague from New Zealand who had worked in our laboratory during his sabbatical contacted me to solicit my opinion on the subject. After much discussion between us, and after I 'retired', we decided to cooperate on a poster-session paper that was presented at the Twenty-Fifth International Symposium on Combustion in 1994[1].

The analysis showed that atmospheric water vapor played the dominant role in infrared absorption, and that any 'greenhouse runaway' for the Earth's temperature should therefore already have occurred long before the last century's increase in atmospheric CO_2. With the ocean's water vapor flux increasing exponentially with temperature, the resultant increase in cloud cover albedo would naturally limit or 'buffer' the system with negative feedback.

The paper also challenged the two 'Greenhouse Catechisms'. The first catechism argues that the in the absence of the 'greenhouse effect', the Earth's temperature would be too cold for human habitation (about -25°C). It is argued that it is the atmosphere that 'keeps the heat in'.

---

[1] Hertzberg, M. and Stott, J. B. 1994, *Greenhouse Warming of the Atmosphere: Constraints on Its Magnitude*, 25th International Symposium on Combustion. Abstracts of Work in Progress Poster Session Presentation, The Combustion Institute, p. 459. (The complete poster session publication is available on request from the author of this publication)

That sets us up for the argument that too much greenhouse effect from too much CO_2 will make the Earth too hot for human habitation. This first catechism will be referred to in a later figure as the 'Cold Earth Fallacy', and it is based on the erroneous assumption that the earth's surface and all the other entities involved in its radiative losses to free space all have unit emissivity. The second catechism has already been discussed: the contention that Venus' high surface temperature is caused by the 'greenhouse effect' of its CO_2 atmosphere.

As fear-mongering hysteria about human-caused global warming grew, and as the Kyoto protocol was promulgated, I felt compelled to get our analysis published more widely. I wrote to Bert Bolin, then head of the IPCC, and submitted our paper to Nature and Science, but they refused to publish it. Who were we to challenge all those sophisticated computer models that were predicting catastrophic warming as a result of human CO_2 emission?

After some correspondence with the editor of Nature, and when it became clear that they were uninterested in publishing the results of our analysis, I felt compelled to candidly express my opinions on the entire question. Here, in a condensed form, is the content of my last letter to the editor of Nature in October, 1994.

I have just reviewed the two articles you referenced... The article by C— is an excellent survey of the complexities involved in the hydro-geological cycle...its emphasis on the necessity of obtaining more data is certainly something with which I agree...Our analysis is certainly consistent with his survey, but our analysis also offers the simplest of models...the radiative equilibrium perspective.

I plead guilty to simplicity...the largest mass and heat capacity in the hydrogeological cycle is in the oceanic component of that cycle, and if one applies Kirchhoff's law to the system, the ocean is in radiative equilibrium with the solar irradiance. The details of the composition of the dry atmosphere are thus of little account in the overall balance since the law is valid for any composition. At its equilibrium temperature, water can accumulate in its deep ocean storage realm to provide a long term "memory" of that equilibrium condition.

The atmosphere is not driven by the short-term 'forcing function' of absorption within the atmosphere's relatively trivial mass, but rather by the long-term 'forcing function' of the memory of the accumulated radiative equilibrium that resides in the ocean. In the intermediate term, the atmosphere is driven by variations in ocean dynamics in accordance with the El Nino phenomenon (i.e. the Pacific Decadal Oscillation). In the longer term, it is driven by variations in solar irradiance associated with variations in the Earth's orbital motion about the Sun in accordance with Milankovitch.

(I clearly neglected to include the variations in the solar cycles and how they might influence cloudiness).

The current 'greenhouse models' such as those referred to in the W— & R— article have it 'backasswards': they drive the oceans with the atmosphere, which is an absurd notion that is contradicted by everything we know about long range weather forecasting!

When I first read your comment that 'Model validation using existing observational data is a fairly standard procedure', my initial reaction was: hurrah, at last someone has made an honest attempt to validate their model. But the euphoria lasted only as long as it took me to read the article in question by W—& R—. There is nothing in that paper that deals with model verification!

There is absolutely nothing in that article that compares the standard greenhouse 'radiation forcing' 'scenarios' or 'projections' with data. The article contains all the standard 'politically correct' projections that have appeared over and over again in the literature.

Over the years, I have done battle with many 'combustion modelers' in considering the question of whose responsibility it is to verify the validity of a proposed model. Was it the responsibility of their readers; was it theirs as formulators of the model; or was it mine as a reviewer and editor? In most cases the modelers seemed satisfied if their model agreed with one observation, or maybe even two if they were allowed to include some 'fudge factors'. I never detected much enthusiasm on their part for searching for a large array of data to test against their models. They seemed happy to get a publication under their belts by proclaiming a model in print, and then leaving it up to everyone else to validate or (heaven forbid) invalidate their models. The literature is now heavily polluted with 'computer experiments' that only serve to corrupt both our thinking and our language.

The situation is far worse with the greenhouse modelers.

The current spate of greenhouse models is motivated in part by the same desire for publication, by the perceived need to create new departments in Universities that will deal with this critical problem of 'global weather change', and by the politics of the environmental movement which encourages the projection of catastrophes...

In the combustion field, the proliferation of unverified models results in limited damage: there is some confusion in thought, and it encourages the illusion that one need no longer do real experiments. There is also some diversion of resources from the real world to the fantasy world of modelers. Nevertheless, there is some educational value in having graduate students learn to handle the conservation laws of energy, mass, and momentum even though they are typically solved for only one dimension, and without buoyancy, and for trivial flows that do not represent real world (turbulent) flow *fields. But in 'global warming', we are talking about real money: an enormous waste of resources in regulating carbon dioxide emissions as we chase the great greenhouse phantom (or dragon if you wish). Is it unreasonable to require that the models on which the 'projections' are based should be validated or invalidated, and that the effort be genuine, nonpolitical, and objective? Should not an alternate model that formulates the problem in terms of radiative equilibrium be considered by the same readership? I have been dismayed to find the arguments I refer to in my paper as the 'first greenhouse catechism' being presented uncritically in first year physics texts and biology courses. As a research scientist and teacher, I feel obligated to do everything I can to correct such misperceptions, and would appreciate NATURE's help in the matter.*

My pleas to *Nature* clearly fell on deaf ears.

But the final defeat came when I was even rejected by my own Unitarian Universalist Association. They were on the way to adopting a resolution on global warming. I tried to present the skeptics position at their General Assembly in Long Beach several years ago, but was not allowed to do so even though I knew more about the subject than anyone else there. I was told that it was 'settled science' and what they wanted was action to curb greenhouse gas emissions. In a workshop at a more recent General Assembly in Salt Lake City, I rose from the audience to present the skeptics viewpoint. But as soon as it became clear that I opposed their position, someone jumped up immediately and grabbed the microphone away from me. No one in the audience defended my right to present my arguments. So much for the fourth principle of the Unitarian Universalist Association:

A free and responsible search for truth and meaning.

Counterattack as Reinforcements Arrive

In 2001, my wife and I took a Nation magazine cruise along the west coast of Mexico. One of the featured speakers during the cruise was their columnist Alexander Cockburn, who is also the co-editor of the magazine, Counterpunch. I sensed from some of his comments that he had serious reservations about the theory of human caused global warming. I spoke to him after one of his talks, indicating that I was a scientist who had been studying the question for several years. He indicated an interest in the result of my studies, so I sent him copies of my 1994 paper, my several letters to the editor, and other correspondence.

After a hiatus of about six years, and out of a clear blue sky, he called me on the telephone to inform me that he was preparing to write series of articles in *Nation* on the subject. I agreed to provide him with scientific advice. The articles appeared in four issues from May 14-June 25, 2007, with letters to the editor and his responses in the June 18 issue. The articles appeared under the intriguing titles; *Is Global Warming a Sin?*, *Who Are the Merchants of Fear?*, *The Greenhousers Strike Back and Strike Out*, and *Dissidents Against Dogma*. After the Climategate scandal broke, he wrote another article that appeared in the Jan 4, 2010 issue entitled *From Nicaea to Copenhagen*. Letters to the editor and responses to that article appeared in the Feb. 8 issue.

Cockburn has received vituperative criticism from environmentalists as a result of that series of articles, and I myself was accused of being a tool of the coal barons. That would be a great surprise to them since I spent most of my career advocating for more stringent safety regulations in their mines.

Martin Hertzberg

The Earth's Radiative Equilibrium

I am exceedingly grateful to Cockburn for his series of articles about global warming and for the discussions we had on the scientific issues. He is one of the few journalists who has exercised due diligence in trying to understand the science. Most others in the 'mainstream media' simply regurgitate the anecdotal, fear-mongering claptrap they are fed by environmental lobbyists without digging any deeper into the totality of the data available or the fundamentals of the science involved.

That interaction with Cockburn encouraged me to revisit, amplify, and update the 1994 poster session paper. The new paper was published in early 2009 in Energy and Environment[2]. I am also grateful to Fred Goldberg, my friend and colleague from Sweden, who was kind enough to review that paper. Fred has been a long time skeptic who has openly challenged the IPCC's conclusions on human caused global warming and even publicly confronted Bert Bolin on the question. Both Cockburn and Goldberg spent a week with us at our residence last year in an impromptu salon discussing the science and politics of the 'global warming/climate change' issue.

Fred presented a spellbinding lecture on the climate history of Scandinavia to my Meteorology class at Colorado Mountain College, and even skied with us at Copper Mountain.

We now proceed to the analysis in that 2009 paper[3].

If one balances the solar input power absorbed by all the Earth's entities involved in the radiative balance between the Earth and the Sun against the power lost by those same entities as they radiate to free space, one obtains an equation for the average

---

[2] Hertzberg, M. 2009, *Earth's Radiative Equilibrium in the Solar Irradiance*, Energy and Environment, Vol 20, No. 1, pp 83-93. Or at www.icecap.us/images/uploads/EE20-1_Hertzberg.pdf
[3] Op cit.

equilibrium temperature of those entities, which shows that the controlling factor is the ratio of their absorptivity to the emissivity.

Their absorptivity is controlled by the fraction of the Sun's radiation that is reflected back to space, which is the Earth's albedo, and is determined mostly by its cloud cover. A high albedo means a low absorptivity, and a low albedo gives a high absorptivity.

Figure 1 is a plot of that average equilibrium temperature for those entities on the Earth in degrees Celsius as a function of the emissivity for four values of the albedo.

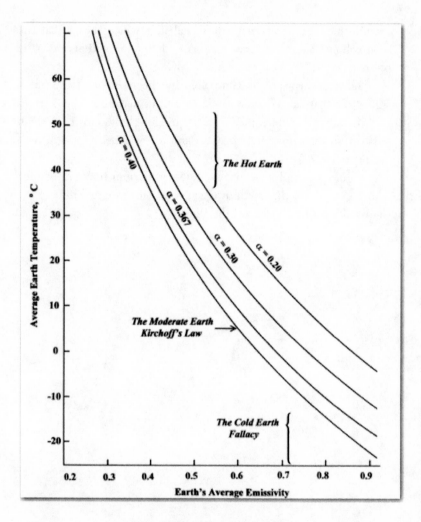

Figure 1: Average Equilibrium Temperature for those
Entities on the Earth in Degrees Celsius as a Function of
the Emissivity for Four Values of the Albedo[4]

Taking the logarithms of the equation for the equilibrium temperature, and taking differentials of the result, allows one to calculate the change in the average temperature of those entities associated with various changes in their emissivity, absorptivity, or the albedo.

That sensitivity curve is plotted in Figure 2 for the current average atmospheric temperature of 291 K, and for an average albedo of 0.30.

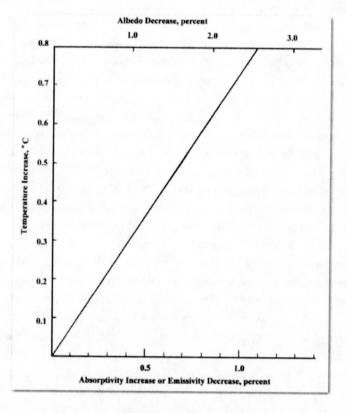

Figure 2: Plot of Albedo vs. Temperature[5]

---

[5] Op cit.

It is at this point that it must be acknowledged that there is considerable uncertainty in determining what 'entities' on the earth are involved in its radiative equilibrium with the Sun and free space. The solar input radiation is absorbed both heterogeneously and homogeneously: heterogeneously at the tops of clouds and at the Earth surface, and homogeneously by the gaseous components of the atmosphere. The same distribution of those absorbers are emitters of the flux that is radiated from the Earth to free space.

Those entities are distributed vertically throughout the atmosphere: from the ocean surfaces at sea level, to the mountains at high altitude, to continental depressions below sea level, and to the upper reaches of the atmosphere at the tops of clouds, and to other particulates suspended in the atmosphere. Those same entities are distributed longitudinally and latitudinally from the equator to the poles. With what measured temperatures are the calculated ones to be compared? Is it reasonable to expect calculated temperatures should be compared only with the air temperatures measured near the Earth's topographic surface?

How representative is such an average surface air temperature of the temperature of the entire mass of the atmosphere involved in the radiative equilibrium process? If the near surface air temperature is not representative, is it realistically possible to measure the average temperature of the entire mass of absorbing and emitting entities with sufficient accuracy to make a meaningful comparison between the data and predictions? One is asking for a definition of the mass of matter that constitutes the Earth's surface, atmosphere, and oceans. How high in altitude should one go in the atmosphere to include it all?

Similarly, how deep in the liquid fluid of the oceans should one go in order to include the mass below the ocean surface that influences the heat and mass transport processes near the ocean surface and in the atmosphere above it? How representative are those near surface temperatures of the average temperature of those vertically distributed yet poorly-defined entities? As difficult as these questions may be, they are nevertheless the ones that need to be answered in order to evaluate the validity of any models purporting to predict future conditions.

This is a formidable task. However, looking at the problem in depth, it may be more realistic to conclude that its resolution may be unattainable given our limited understanding of the complex processes involved, and the lack of data available for the current thermodynamic state of those entities.

Nevertheless, despite those complexities, we continue this analysis by making the reasonable assumption that any changes in the average temperature of those entities will be reflected in similar changes in the average atmospheric temperature near the Earth's surface, as measured by the meteorological network of surface stations or from satellite observations. Those measured temperature changes as reported by the IPCC over the last century[6] (3) are as follows:

1910-1940, an increase of 0.5°C
1940-1970, a decrease of 0.2°C
1970-2000, an increase of 0.5°C

As can be seen from Figure 2, those increases of 0.5°C for the two thirty year spans from 1920 to 1940 and from 1970 to 2000 correspond to a relatively small decrease of only 1.5 percent in the Earth's albedo. The observed decrease in temperature of

---

[6] Klyashtorin, L. B. and Lyubushin, A. A., *On the Coherence Between Dynamics of World fuel consumption and Global Temperature Anomaly*, Energy and Environment, 2003, Vol. 14, No. 6 pp 773-782, Fig. 1. Also at the National Climate Data Center, Global Surface Temperature Anomalies, 2007, on the web at
www.ncdc.noaa.gov/oa/climate/research/anomalies/anomalies.html

0.2°C from 1940 to 1970 corresponds to an albedo increase of only 0.5 percent.

Thus those modest changes in temperature are readily explained in terms of minor changes in albedo brought about by small changes in cloudiness. Svensmark[7] [8] has shown that the Earth's cloud cover underwent a modulation in phase with the cosmic ray flux during the last solar cycle. His suggested mechanism for that correlation involves a decrease in cosmic ray flux during high solar activity, when the 'solar wind' and magnetic activity shield the Earth from cosmic rays. The reduced incidence of cosmic rays results in the absence of adequate nucleating agents for cloud formation, a decrease in the Earth's albedo, a corresponding increase in absorptivity, and hence a heating of the Earth.

The opposite occurs during low solar activity, when the cosmic ray flux into the Earth's atmosphere is high, nucleating agents are plentiful, and cloudiness increases the albedo. This results in a decrease in absorptivity and hence a cooling of the Earth. The analysis summarized earlier from Figure 2 supports the Svensmark mechanism as the causes of the 20th Century fluctuations in the average Earth temperature. As Figure 2 shows, relatively modest changes of only a few percent in the Earth's albedo are sufficient to account for the observed temperature changes of that century. Those are precisely the magnitudes of the changes in cloudiness that are observed by Svensmark to vary in phase with the variations in solar activity.

---

[7] Svensmark, H. 2000, *Cosmic Rays and the Earth's Climate*, Space Science Reviews, Vol 93, pp 155-166

[8] Svensmark, H. and N. Calder 2007, *The Chilling Stars: A New Theory of Climate Change*, Icon Books Ltd., Cambridge, 249 pp.

Thus, except for the influence of cloud albedo, no assumptions are needed regarding the detailed composition of the atmosphere in order to explain the observed modest variations in 20th Century temperatures of the Earth's atmosphere.

This analysis supports the earlier conclusion that it is implausible to expect that small changes in the concentration of any minor atmospheric constituent such as carbon dioxide can significantly influence the radiative equilibrium between the Sun, the Earth, and free space.

Puff, the Magic Sky Dragon is gone

At the present time, global warming skeptics/realists/deniers fall into two camps. The first camp believes that the greenhouse gas warming phenomenon is real, but the degree of warming from the recent increases in atmospheric CO_2 concentrations is trivial. The second camp denies the very existence of the greenhouse effect— arguing that it is totally devoid of physical reality and that as traditionally defined, violates the laws of thermodynamics.

We here attempt to resolve the question by idealizing the radiative transport processes between the Earth's surface, its atmosphere and free space in the absence of any solar input radiation.

As indicated earlier, the problem of obtaining accurate absorptivity-to-emissivity ratios for all the entities on the Earth that participate in the radiative balance is a formidable task. It is highly unlikely that any proposed model contains a realistic ratio for the entire globe over a long enough time scale. But, even if those quantities were precisely known, the resultant temperature structure of the system of entities cannot be determined until all other energy transfer processes and forces are included in the model.

Those other processes involve conduction, natural convection, forced convection (advection to meteorologists) in

179

both the atmosphere and the oceans, endothermic evaporation from the oceans and land, exothermic condensation of water vapor in the atmosphere, and their accompanying mass transport processes, and finally, the intractable problem of turbulence. To those processes must be added the buoyancy force couple, the Coriolis force, and the tidal forces.

Thus, even if the radiative processes were precisely known, all the other processes just cited would have to be included in order to predict the temperature structure of all the Earth's entities. The complexity of the problem boggles the mind and has frustrated forecasting meteorologist for decades.

Instead, let's consider reversing the process. What can be learned from using the known thermal structure of the Earth's surface and its atmosphere, and then inferring the radiative transport processes that must accompany that structure? This analysis is taken from a paper entitled *The Nighttime Radiative Transport between the Earth's Surface, its Atmosphere, and Free Space* that has recently been submitted for publication in Energy and Environment. The analysis reflects the radiative fluxes for nighttime conditions—but they also are also present during daytime conditions when they must be subtracted from the input solar fluxes in order to obtain the net amount that heats the Earth.

The Earth's surface, its atmosphere, and free space, are approximated as concentric spherical surfaces whose radii are much larger than the distance between them and whose average temperatures, emissivities, and absorptivities are known. The Earth's surface entities are taken to be at its average temperature, its average emissivity and its average absorptivity. The gaseous atmosphere *without* clouds to begin with, is approximated as a partially-absorbing, partially-transparent, non-reflective glass-like plate at a colder average temperature with its average absorptivity and its average emissivity.

The gaseous atmosphere is condensed into a thin glass plate whose average temperature is taken as the temperature of the

'Standard Atmosphere' half way up at the 500mb surface. When all is said and done, one obtains the following result...the net amount of infrared radiation absorbed by the colder atmosphere above from that emitted by the warmer atmosphere below is 25 W/m^2. The infrared radiation lost to free space from the atmosphere is 46 W/m^2. The infrared radiation lost from the Earth's surface to free space that is transmitted through the atmosphere is 228 W/m^2.

Thus it is clear that the atmosphere helps to cool the Earth—atmosphere system, and that in the absence of clouds, it accounts for some 17% of the radiant energy flux that the system as a whole loses to free space.

The general correctness of this picture is clearly confirmed by the fact that direct meteorological soundings of the atmospheric lapse rate show that both the Earth's surface and the atmosphere both cool during night-time hours, albeit at different rates because of their different emissivities.

It should be noted that nowhere in this balance is there a so-called 'greenhouse effect' in which the atmosphere supplies any net radiant energy that is absorbed by the Earth. Under these assumptions for the thermal structure, the flow of radiant energy from both the earth's surface and its atmosphere is entirely outward toward free space.

In the presence of clouds covering on the average some 33 % of the Earth's surface, the 'glass plate' atmosphere becomes partially reflective. For that cloudy atmosphere, the radiation from the atmosphere to free space increases to about 106 W/m^2 and the radiation lost from the surface to free space is decreased to 153 W/m^2.

With clouds, the atmosphere now accounts for some 41 % of the total radiant flux lost to free space. The physical effect of that radiant loss from clouds to free space is apparent from the fact that thunderstorm activity tends to maximize after sundown because of radiation from the tops of clouds. That radiation loss results in marked cooling of those cloud tops which steepens the temperature lapse rate, increasing the instability of the cloudy atmosphere and thus increasing thunderstorm activity. As was the case for the cloudless atmosphere, for the cloudy atmosphere, the so-called 'greenhouse effect' is nowhere to be found in the radiative balance. All the radiant flux is outward toward free space.

There is only one exception where one can find a net radiant flux from the atmosphere to the Earth's surface, and that occurs during atmospheric inversion conditions. But even in the extreme case in which the surface temperature and the atmosphere's temperature are reversed, the radiant power lost to free space from the atmosphere is a factor of five greater than the power radiated toward the surface from the warmer atmosphere. Inversion conditions are thus the only case in which the so-called 'greenhouse effect' can possibly have any form of physical reality. But, of course, that is not how the greenhouse effect is traditionally defined by global warming modelers.

Such inversion conditions, however, are present over a small fraction of the Earth's surface for limited periods of time, and since the recent increases in atmospheric CO_2 concentrations have virtually no effect on the atmosphere's total emissivity, the effect of those CO_2 increases on the overall radiative flux balance is essentially nil.

The Legend of the Sky Dragon and Its Mythmakers

There is a simple way to tell the difference between propagandists and scientists. If scientists have a theory they search diligently for data that might actually contradict the theory so that they can fully test its validity or refine it. Propagandists, on the other hand, carefully select only the data that might agree with their theory and dutifully ignore any data that disagrees with it.

One of the best examples of the contrast between propagandists and scientists comes from the way the human caused global warming advocates handle the Vostok ice core data from Antarctica[9].

The data span the last 420,000 years, and they show some four Glacial Coolings with average temperatures some 6 to 8°C below current values and five Interglacial Warming periods with temperatures some 2 to 4°C above current values. The last warming period in the data is the current one that started some 15,000 to 20,000 years ago. The data show a remarkably good correlation between long term variations in temperature and atmospheric CO_2 concentrations. Atmospheric CO_2 concentrations are at a minimum during the end of Glacial Coolings when temperatures are at a minimum. Atmospheric CO_2 concentrations are at a maximum when temperatures are at a maximum at the end of Interglacial Warmings. Al Gore, in his movie and his book, *An Inconvenient Truth*, shows the Vostok data, and uses it to argue that the data prove that high atmospheric CO_2 concentrations cause global warming.

Is that an objective evaluation of the Vostok data? Let's look at what Gore failed to mention. First, the correlation between temperature and CO_2 has been going on for about half a million years, long before any significant human production of CO_2, which began only about 150 years ago. Thus, it is reasonable to argue that the current increase in CO_2 during our current Interglacial Warming, which has been going on for the last

---

[9] Petit, J. R. et al 1999, *Climate and Atmospheric History of the Past 420,000 Years from the Vostok Ice Core, Antarctica*, Nature, Vol 399, pp 429-436

15,000-20,000 years, is merely the continuation of a natural process that has nothing whatever to do with human activity.

Gore also fails to ask the most logical question: where did all that CO_2 come from during those past warming periods when the human production of CO_2 was virtually nonexistent? The answer is apparent to knowledgeable scientists: from the same place that the current increase is coming from, from the oceans. The amount of CO_2 dissolved in the oceans is some fifty times greater than the amount in the atmosphere. As oceans warm for whatever reason, some of their dissolved CO_2 is emitted into the atmosphere, just as your soda pop goes flat and loses its dissolved CO_2 as it warms to room temperature even as you pour it into the warmer glass. As oceans cool, CO_2 from the atmosphere dissolves back into the oceans, just as soda pop is made by injecting CO_2 into cold water.

But the real 'clincher' that separates the scientists from the propagandists comes from the most-significant fact Gore fails to mention. The same Vostok data show that changes in temperature always precede the changes in atmospheric CO_2 by about 500-1500 years.

The temperature increases or decreases come first, and it is only after 500-1500 years that the CO_2 follows. Figure 3 shows the data from the termination of the last Glacial Cooling (Major Glaciation) that ended some 15,000-20,000 years ago through the current Interglacial Warming of today. The four instances where the temperature changes precede the CO_2 curve are clearly shown. All the Vostok data going back some 420,000 years show exactly the same behavior. Any objective scientist looking at that data would conclude that it is the warming that is causing the CO_2 increases, not the other way around as Gore claimed.

I am indebted to Guy Leblanc Smith for granting permission to use Figure 3 below as it was published on Viv Forbes' web-site www.carbon-sense.com.

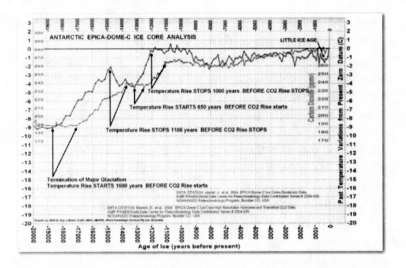

Figure 3: Antarctic EPICA-C Ice Core Analysis[10]

It is even more revealing to see how the advocates of the human-caused global warming theory handle this 'clincher' of the argument. It is generally agreed that the Vostok cycles of Glacial Coolings and Interglacial Warmings are driven by changes in the parameters of the Earth's orbital motion about the Sun and its orientation with respect to that orbit; namely, changes in the ellipticity of its orbit, changes in its obliquity (tilt relative to its orbital plane), and the precession of its axis of rotation.

These changes are referred to as the Milankovitch cycles, and even the human caused global warming advocates agree that those cycles 'trigger' the temperature variations. But the human-caused global warming advocates present the following *ad hoc* contrivance to justify their greenhouse effect theory. The Milankovitch cycles, they say, are 'weak' forcings that start the process of Interglacial Warming, but once the oceans begin to release some of their CO_2

---

[10] From www.carbon-sense.com

after 500-1500 years, then the 'strong' forcing of 'greenhouse warming' takes over to accelerate the warming.

That argument is the best example of how propagandists carefully select data that agrees with their theory as they dutifully ignore data that disagrees with it. One need not go any further than the next Glacial Cooling to expose that fraudulent argument for the artificial contrivance that it really is. Pray tell us then, we Slayers of the Sky Dragon ask, what causes the next Glacial Cooling? How can it possibly begin when the CO_2 concentration, their 'strong' forcing, is at its maximum? How can the 'weak' Milankovitch cooling effect possibly overcome that 'strong' forcing of the greenhouse effect heating when the CO_2 concentration is still at its maximum value at the peak of the Interglacial Warming?

The global warmers thus find themselves stuck way out on a limb with that contrived argument. They are stuck there in an everlasting Glacial Warming, with no way to begin the next Glacial Cooling that the data show.

But one has to be sorry for Gore and his friends, for after all, they are in the global warming business. Global cooling is clearly someone else's job!

In my 1994 paper, it was concluded that the unverified models used by the IPCC did not realistically represent the forces that determine the temperature of the Earth and its atmosphere, and that it would be absurd to base public policy decisions on them. Regrettably, what was then merely 'absurd' has today turned into something more sinister.

Models have been developed that try to validate the existence of an intensifying 'greenhouse effect' driven by very modest changes in the concentration of the minor constituent, CO_2, even though its absorption of the Earth's infrared radiation emitted to free space is already near saturation. Those models continue to be developed even though as shown here, and as shown much earlier, the 'greenhouse effect' has long been known to be devoid of physical reality[11] [12]. When those models were criticized for their omission of clouds, the modelers included water vapor, but in the form of a positive feedback. That way the models could magnify the trivial effect of increasing CO_2 concentrations, and thus 'tweak' them in the direction the modelers wanted them to go. In doing so, they ignored the overwhelming evidence that water vapor feedback in the form of clouds is negative.

Even after their models were shown to be faulty, they continued to use them to make predictions, which were then touted as the equivalent of actual data, and public policy decisions were then made, and continue to be made, on the basis of those models.

Overall, such disingenuous behavior, and the acceptance of such behavior by some scientific journals, professional societies, and government agencies, both national and international, essentially amounts to scientific malfeasance on a grand scale.

The implementation of policies based on the acceptance of such malfeasance will continue to have damaging effect to both science and the public welfare.

---

[11] Wood, R. W. 1909, *Note on the Theory of the Greenhouse*, Philosophical Magazine, Vol. 17, pp 219-320

[12] Gerlich, G. and R. D. Tscheuschner 2009, *Falsification of the Atmospheric CO_2 Greenhouse Effects Within the Frame of Physics*, International Journal of Modern Physics B, Vol. 23, No. 3, pp 275-364. Available at http://arxiv.org/PS_cache/arxiv/pdf/0707/0707.1164v4.pdf

This chapter provides additional and alternative explanations regarding the erroneous concepts of an atmospheric greenhouse effect caused by greenhouse gases.

Chapter 13

The Bigger Picture
by Hans Schreuder

THE UN'S IPCC bases its dire forecasts on nothing more than computer models that represent the earth as a flat disk bathed in a constant twenty-four-hour haze of sunlight without north and south poles and with few clouds and thus without any relationship to the real planet we all live on.

Despite much rhetoric and research over the past decades, there is still not a single piece of actual evidence that the now-maligned carbon dioxide molecule causes global warming (or 'climate change' for that matter).

To over 40,000 fellow scientists from around the world—and to me—this is no surprise, for no such evidence can ever be found.

Carbon dioxide (CO_2), at less than 400 parts per million by volume, does not and cannot influence either the atmospheric temperature or the climate in any measurable way. Only laboratory experiments with heat lamps can make carbon dioxide do what climate change proponents want it to

189

do: warm the flasks that contain CO_2. Yet this is not principally how the open atmosphere gets heated and no laboratory experiment can mimic actual air dynamics or be extrapolated to represent them.

The earth's air hugs the surface like a thin shell which is completely encapsulated by a perfect thermal insulator: the vacuum of space. Earth does not need a 'blanket of greenhouse gases' to keep it warm or protect it from the cold of space. The vacuum of space is the best possible insulator we could wish for. A widely-held concept that space is cold is very off course. Space is not cold in the same way that we feel cold; it has no temperature of itself. It is a vacuum and a vacuum has no temperature. Only matter can have a temperature and in a vacuum, there is as good as no matter.

Scientists discovered what they call Cosmic Background Radiation, seemingly indicating that 'space' has a temperature of 3K (some minus 270°C; minus 454°F), but this is a misleading concept. Only the very few molecules of matter within that vacuum of space exhibit that actual temperature of only 3K.

The vacuum around that matter cannot have a temperature; it is a vacuum after all and a vacuum, by definition, does not contain anything. In other words, it is a misconception that the earth's temperature needs insulation to begin with, let alone that a trace gas at barely 400 parts per million by volume is providing this insulation.

The as yet poorly understood adiabatic process of pressure increase or decrease by way of greater or lesser altitude supposedly generates enough heat to keep the various regions of our earth at a near-constant temperature whilst near-constant solar radiation provides the extra heat and energy for life as we know it. There are irregularities with the adiabatic process that we do not yet fully understand and there is not one single continuous temperature reduction with ever increasing altitude. As shown in Figure 1, the earth is not alone in this.

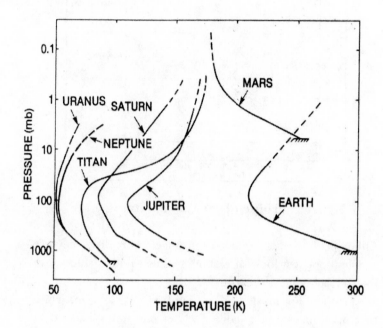

Figure 1: Temperature vs. Pressure for the Planets[1]

In greater detail, as shown in Figure 2, our earthly atmosphere displays even more variations of temperature with ever increasing altitude.

---

[1]

http://lasp.colorado.edu/~bagenal/3720/CLASS14/AllPlanetsT. jpg

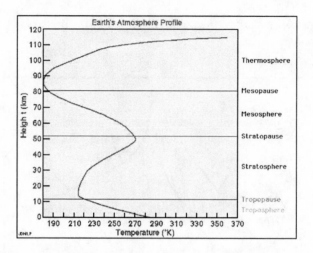

Figure 2: Profile of the Earth's Atmosphere[2]

Although the issues involved are hugely complex, they are simple if one just looks at carbon dioxide's potential to warm the atmosphere or the earth.

First and foremost, air itself (oxygen and nitrogen, which together make up as good as 99% of our atmosphere) does not respond well to the electromagnetic radiation which CO_2 reacts to.

Consider a microwave oven for instance, where the interior's air is not warmed by the microwaves but by the heated food instead. The food is heated by the microwaves and then the food warms the air by conduction and convection. This roughly simulates how the surface of the earth warms the swirling air that comes into contact with it. Yet the IPCC has it that energy radiated by the earth is re-radiated back by 'greenhouse gases' which make the system

---

[2]

http://eesc.columbia.edu/courses/ees/slides/climate/atmprofile.gif

ever warmer. This second-hand infrared energy supposedly causes a warming of the troposphere (that's the lowest part of our atmosphere and within which we live), as depicted in the UN IPCC graphic of Figure 3.

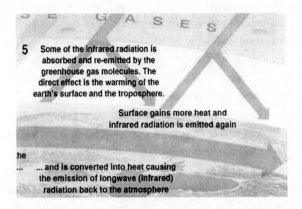

Figure 3: The IPCC View of the Effect of Greenhouse Gases[3]

Without a cause, however, there can be no effect. This is why the predicted greenhouse tropospheric 'hot spot' has never been found, quite the opposite in fact. But not only is the hot spot not there, it cannot be there. Without a cause there can be no effect[4].

As per that same IPCC graphic, re-radiated infrared energy is also supposed to warm the earth. In reality, energy that is re-radiated by a molecule spreads out in three

[3]

http://www.appinsys.com/GlobalWarming/GW_PART5_GREE NHOUSEGAS_files/image001.jpg
[4]

http://scienceandpublicpolicy.org/monckton/greenhouse_warmi ng_what_green house_warming_.html dimensions. Thus only about 35% (at best) can be directed back to where it came from—the rest of it goes sideways or upwards.

But, critically important, re-radiated energy cannot make a heat source any warmer than it was in the first place. If it could, we would have found the holy grail of energy, a perpetuum mobile whereby more energy is extracted than what goes in.

If reflecting heat back to a heat source raises its temperature, then just reflecting it again will raise its temperature even more, and so on, till a one watt input generates a billion watts of power. That's clearly impossible. Yet this child's version of science has charmed much of the world into uncritical belief.

Secondly and of equal importance is the fact that human activities constitute about 3% of the yearly emissions total. More than 98% of this total is absorbed within a year (thus contradicting the long residence claim). Since 1.5% is left over, which is recorded as the increase of atmospheric CO_2, the human contribution is only 3% of this 1.5%. This means that, as a maximum, only some 14 PPMV (Parts Per Million by Volume) of the increased levels of carbon dioxide can be ascribed to human activities, as indicated by figures provided by the U.S. Department of Energy and the IPCC[5].

---

[5]

ftp://ftp.eia.doe.gov/pub/oiaf/1605/cdrom/pdf/ggrpt/057304.pdf—page 6 (page 26 within the PDF).

The Bigger Picture

Table 3. Global Natural and Anthropogenic Sources and Absorption of Greenhouse Gases in the 1990s

Gas	Sources			Absorption	Annual Increase in Gas in the Atmosphere
	Natural	Human-Made	Total		
Carbon Dioxide (Million Metric Tons of Gas)[a]	770,000	23,100	793,100	781,400	11,700
Methane (Million Metric Tons of Gas)[b]	239	359	598	576	22
Nitrous Oxide (Million Metric Tons of Gas)[c]	9.5	6.9	16.4	12.6	3.8

Energy Information Administration
Office of Integrated Analysis and Forecasting
U.S. Department of Energy

Source: Intergovernmental Panel on Climate Change, Climate Change 2001: The Scientific Basis
(Cambridge, UK: Cambridge University Press, 2001).

Figure 4: Sources and Absorption of Greenhouse Gases

Third is the inconvenient fact that the world hasn't been warming for more than a decade now (as of 2010), despite a steady and ever-climbing carbon dioxide level, proof enough by itself that no influence over global temperatures is to be gained from extra atmospheric carbon dioxide.

Actual observed evidence needs to be put on the table, not computer model outputs or presumptively-inferred evidence. Glaciers are not melting in alarming fashion, the Greenland icecap is not collapsing and the Arctic is not about to become ice-free. Neither is the Antarctic melting away and sea levels are *not* rising any faster than they have done for the past 11,000 years—there is simply no irrefutable evidence indicating a warming role for atmospheric carbon dioxide.

Any and all alarmist predictions and observations have been decisively disproved over the past decade, whilst global temperatures have been going down rapidly instead of ever up as had been so widely predicted by the constant tweaking of climate models.

Based on the behavior of the one and only true climate driver, our sun, the western governments would be better

195

advised to prepare for longer, colder winters and shorter growing seasons for many decades to come.

A favorite expression used by climate alarmists and skeptics alike is the blanket effect. Let's examine that in closer detail: a blanket returns your own heat to yourself and that's why you become warmer?

Or is it perhaps that a blanket prevents convection and thus your body can not freely dispose of its generated heat as in a real greenhouse with glass panes or plastic sheeting or metal sheeting or even a wooden shed. Stop or hinder convective heat loss and bingo, the cooling process is interrupted. No extra heat is generated, if only. It just takes longer for the same amount of energy to disperse itself. That's how a thermos flask works after all—despite the best possible 're-radiation' of the same energy, the contents of the flask will cool down if you started with a hot substance in it.

It's the same with our open-to-space atmosphere. Our atmosphere is surrounded by the vacuum of space exactly the same as a thermos flask (vacuum flask). Remember that space is not cold. Space has no temperature—there is not enough matter in the vacuum of space for it to have a 'temperature'.

Yes, there is background radiation which indicates the vacuum of space is only 3K—that's mighty cold. Yes, the odd bit of matter that comprises the vacuum of space will indeed by at 3K, but that will not make the vacuum of space in its vastness at 3K as well—how could it?! It's a vacuum—how can nothing have a temperature?

Nothing makes the atmosphere 'warmer than it would be'. The insulation of the vacuum of space in which earth and its atmosphere finds itself acts like the most perfect insulator, just like the vacuum flask. Water vapor slows down the cooling rate at night provided there is sufficient water vapor in the atmosphere; it does not make the night 'warmer', it merely keeps it warm longer. Even in the hottest tropics the end of the night will be cooler than the end of the previous day. Water vapor has a huge capacity for latent heat (hidden heat) and that's the only reason that the tropics are so much "warmer" at night than more temperate zones.

Without an atmosphere, earth would mimic our moon: very hot during the day, very cold during the night[6]. Keep in mind: the same water vapor makes the tropics cooler during the day than it would be without the vapor. Just think of a dry desert and a tropical region at the exact same latitude (in Southern Africa for instance).

Dry desert: hot during the day, cold during the night.

Tropical region: cooler than the desert during the day, not as cold during the night.

The only difference: water vapor content of the atmosphere. So you have a cooling effect during insolation and a blanket effect during the night. But at no stage is the night 'warmer' than the sun could have made in during the preceding day. Where would that extra energy have come from? And of course there is no way that the day is 'warmer than it would be' due to atmospheric gases: *instead, they act as coolants*[7].

But, mathematically, you could argue that the 'average temperature' is higher due to water vapor, but that is cheating with a formula. The maximum temperature is lower than it would be during the day yet higher than it would be at night.

To average those two points is meaningless in understanding what is going on.

---

[6] http://www.tech-know.eu/uploads/Greenhouse_Effect_on_the_Moon.pdf

[7] *The Atmosphere acts an air conditioner cooling/warming the Earth by combination of thermodynamics and radiation*
http://claesjohnson.blogspot.com/2010/08/energy-budgets-without-backradia tion.html

Chapter 14

Sun heats Earth, Earth heats Atmosphere
by Hans Schreuder

AFTER ALL IS said and done, it will be found that carbon dioxide does not and cannot affect either the global temperature or climate change. Carbon dioxide has no climate forcing effect and is not a greenhouse gas and, for that matter, neither is water vapor.

> *Our understanding of the natural world does not progress through the straight forward accumulation of facts because most scientists tend to gravitate to the established popular consensus also known as the established paradigm. Thomas Kuhn describes the development of scientific paradigms as comprising three stages: prescience, normal science and revolutionary science when there is a crisis in the current consensus. When it comes to the science of climate change, we are probably already in the revolution state[1].*
> —Jennifer Marohasy, 2009.

---

[1] http://www.sott.net/articles/show/183475-Jennifer-Marohasy-Commentary-on-Ferene-Miskolczi-s-Atmospheric-Model

The only worthwhile source of warmth for planet earth is our Sun, warming all of the land and all of the seas, which then warm the atmosphere—not the other way around; the atmosphere does not warm the earth, other than during short-term exceptional weather conditions such as the Sirocco winds over the Canary Islands.

> *To understand heat transfer we have to keep in mind that heat is not a substance, but energy that flows from one system toward other systems with lower density of energy.*[2]
> —Nasif Nahle

Volcanoes add a small amount of heat locally as and when they erupt and sometimes may cause temporary global cooling until the ash and other material has settled back to earth. Erupting underwater volcanoes will add some warmth to the sea, but in the bigger picture, it is only the sun that adds global warmth to our planet.

The atmosphere is mostly warmed up from the heat that radiates off the surface of the earth. During the day, the atmosphere in fact helps to cool the earth and, depending where on earth you are, during the night the atmosphere will either continue to cool the earth (at the poles and in dry deserts) or keep the earth warm (at the equator). Water vapor helps to maintain some of the daytime warmth during the nighttime—the greater the humidity, the greater the capacity of the atmosphere to maintain temperature.

At no stage though does water vapor *add* warmth to the atmosphere and neither does carbon dioxide—only in closed

---

[2]

http://www.biocab.org/Heat_Stored_by_Atmospheric_Gases.html

test flasks in a laboratory, but under no circumstances in the open atmosphere in which we all live.

Before discussing the issue of man-made global warming (AGW) or the man-made climate change, one central definition has to be stated quite clearly.

The so-called greenhouse effect of the atmosphere is most commonly explained as follows:

> *The heating effect exerted by the atmosphere upon the Earth because certain trace gases in the atmosphere (water vapor, carbon dioxide, etc.) absorb and reemit infrared radiation. [...] The component that is radiated downward warms Earth's surface more than would occur if only the direct sunlight were absorbed.*
>
> *The magnitude of this enhanced warming is the greenhouse effect. Earth's annual mean surface temperature of 15°C is 33°C higher as a result of the greenhouse effect...*[3]
>
> —American Meteorology Society

The above definition is the accepted one by climate alarmists and climate realists alike and is the one that is referred to throughout this chapter. That definition is the 'settled science' heralded by the UN IPCC.

That definition is 100% wrong on all counts.

> *We would be mistaken if we were to think that the change of temperature was caused by CO_2 when, in reality, it was the Sun that heated up the soil. Carbon dioxide only*

---

[3]

http://amsglossary.allenpress.com/glossary/search?id=greenhous
e-effect1

interfered with the energy emitted by the soil and absorbed a small amount of that radiation (0.0786 Joules), but carbon dioxide did not cause any warming. Please never forget two important points: the first is that carbon dioxide is not a source of heat, and the second is that the main source of warming for the Earth is the Sun.[4]
—Nasif Nahle

It started with a genuine concern by senior scientists in Europe and the USA that if uncontrolled, increasing emissions of carbon dioxide and other gases into the atmosphere from burning fossil fuels, mainly coal, could have serious consequences. It is also very important to note that global climate models are unable to produce an output that is verifiable. In other words the output can neither be proved nor disproved. What grounds do those who use these models have to refute observations made by others to the effect that there is no believable evidence of the postulated dramatic adverse changes produced by the models?[5]
—Professor Will Alexander

Throughout the last decade, supporters of the idea of an anthropogenic global warming (AGW) or the impact of an anthropogenic "greenhouse" effect on climate (IAGEC) have been insisting on an erroneous concept of the emission of energy from the atmosphere towards the surface. The AGWIAGEC assumption states that half of the energy absorbed by atmospheric gases, especially

[4]

http://www.biocab.org/Heat_Stored_by_Atmospheric_Gases.html
[5] http://climaterealists.com/index.php?id=3162

carbon dioxide, is reemitted back towards the surface heating it up. This solitary assumption is fallacious when considered in light of real natural processes.[6]
—Nasif Nahle

If there was strong evidence of undesirable changes, then the whole climate change issue would have been resolved long ago. The tragedy is that there is a world-wide policy in the opposite direction. Not only has the observation theory route been avoided, but climate change scientists and their organizations have adopted a policy of deliberately denigrating all those who practice it. Why are they following this thoroughly unethical and unscientific procedure? ...after 20 years of massive international effort (the overwhelming consensus), climate change scientists have still to produce solid, verifiable evidence of the consequences of human activities. They have been unable to proceed beyond claims that climate change will result in the 'intensification of the hydrological cycle' for which there is no scientifically believable evidence. Not only do our studies completely negate the claims made by climate change scientists, but we can demonstrate with a high degree of assurance that all the proposed measures to limit greenhouse gas emissions will be an exercise in futility.[7]
—Professor Will Alexander

...atmospheric gases do not cause any warming of the surface given that induced emission prevails over

---

[6]

http://www.biocab.org/Heat_Stored_by_Atmospheric_Gases.html
[7] http://climaterealists.com/index.php?id=3162

spontaneous emission. During daytime, solar irradiance induces air molecules to emit photons towards the surface; however, the load of Short Wave Radiation (SWR) absorbed by molecules in the atmosphere is exceptionally low, while the load of Long Wave Radiation (LWR) emitted from the surface and absorbed by the atmosphere is high and so leads to an upwelling induced emission of photons which follows the outgoing trajectory of the photon stream, from lower atmospheric layers to higher atmospheric layers, and finally towards outer space. The warming effect (misnamed "the greenhouse effect") of Earth is due to the oceans, the ground surface and subsurface materials. Atmospheric gases act only as conveyors of heat.[8]
—Nasif Nahle

It is human arrogance to think that we can control climate, a process that transfers huge amounts of energy. Once we control the smaller amount of energy transferred by volcanoes and earthquakes, then we can try to control climate.

Until then, climate politics is just a load of ideological hot air.

To argue that human additions to atmospheric CO_2, a trace gas in the atmosphere, changes climate requires an abandonment of all we know about history, archaeology, geology, solar physics, chemistry and astronomy. We ignore history at our peril.

---

[8]

http://www.biocab.org/Heat_Stored_by_Atmospheric_Gases.html

Sun Heats Earth, Earth Heats Atmosphere

*I await the establishment of a Stalinist-type Truth
and Retribution Commission to try me for my crimes
against the established order and politicized science.[9]*
—Professor Ian Plimer

*The Atmosphere acts an air conditioner cooling/warming
the Earth by combination of thermodynamics and
radiation.[10]*
—Professor Claes Johnson

To conclude this chapter, it is necessary to understand that
the underlying drive for control over the use of energy is
based on the principles set out in the United Nation's Agenda
21[11] as well as two other relevant agendas[12] [13].

When the idea of blaming carbon dioxide came to be
understood by those who wished to wield their control over
global affairs, the wheels of political manipulation were set in
motion via the UNFCCC.

All Western governments subscribed to these ideals
without understanding the deeper meaning of the hidden
agendas and lured by the promise of huge subsidies, taxation
and green job creation schemes.

---

[9] Ian Plimer, Professor of Earth and Environmental Sciences at the
University of Adelaide and Emeritus Professor of Earth Sciences at
the University of Melbourne, author of *Heaven and Earth—Global
Warming, the Missing Science* Connor Court (2009)
[10] http://claesjohnson.blogspot.com/2010/08/energy-budgets-
without-backradiation.html
[11] http://www.un.org/esa/sustdev/documents/agenda21/
[12] http://www.mdgmonitor.org/index.cfm
[13] http://www.globio.info/

As a final word on the matter of greenhouse gases and the greenhouse effect, I quote from the most elaborate and accurate scientific paper on the subject:

The atmospheric greenhouse effect, an idea that many authors trace back to the traditional works of Fourier (1824), Tyndall (1861), and Arrhenius (1896), and which is still supported in global climatology, essentially describes a fictitious mechanism, in which a planetary atmosphere acts as a heat pump driven by an environment that is radiatively interacting with but radiatively equilibrated to the atmospheric system. According to the second law of thermodynamics, such a planetary machine can never exist. Nevertheless, in almost all texts of global climatology and in a widespread secondary literature, it is taken for granted that such a mechanism is real and stands on a firm scientific foundation. In this paper, the popular conjecture is analyzed and the underlying physical principles are clarified. By showing that (a) there are no common physical laws between the warming phenomenon in glass houses and the fictitious atmospheric greenhouse effects, (b) there are no calculations to determine an average surface temperature of a planet, (c) the frequently mentioned difference of 33 degrees C is a meaningless number calculated wrongly, (d) the formulas of cavity radiation are used inappropriately, (e) the assumption of a radiative balance is unphysical, (f) thermal conductivity and friction must not be set to zero, the atmospheric greenhouse conjecture is falsified.[14]

—Gerhard Gerlich and Ralf D. Tscheuschner

---

[14] http://arxiv.org/PS_cache/arxiv/pdf/0707/0707.1161v4.pdf

Why Carbon Dioxide is not a pollutant and there can be no temperature-increasing greenhouse effect in our open atmosphere.

Chapter 15

Sun Heats Earth, Clearing Carbon Dioxide of Blame
by Hans Schreuder

Summary

THE IMPORTANCE OF this chapter lies in the fact that we all need to rapidly conclude that any and all hype about mankind's carbon dioxide emissions is based on the incorrect application of science. Despite comments by Lord Stern (It's even worse than I thought[1]) and others, there is no greenhouse effect as described in UN IPCC explanations—and carbon dioxide has nil effect on the global climate and does not cause climate change in any way, shape or form.

This chapter will go against all the established interpretations, including those of many skeptical scientists, yet is based entirely upon the proper application of scientific

---

[1] http://www.independent.co.uk/environment/climate-change/lord-stern-on-global-warming-its-even-worse-than-i-thought-1643957.html

principles, especially those of observation-based evidence, none of which has yet been presented to cast doubt, in even the most circumstantial manner, upon the opposite of what is presented to you here. And that's before we take this statement into account:

> As the glaciological and tree ring evidence shows, climate change is a natural phenomenon that has occurred many times in the past, both with the magnitude as well as with the time rate of temperature change that have occurred in the recent decades.
>
> The following facts prove that the recent global warming is not man-made but is a natural phenomenon.[2]
> —Dr. Gerhard Löbert, Munich. Physicist. Recipient of The Needle of Honor of German Aeronautics.

A lot of obscurantism has been thrown at the nature of radiant energy in order to make the weird propositions of greenhouse gas theory seem plausible. The unalterably downward flow of thermal energy is the very essence of the second law of thermodynamics, for instance, but academics will try to argue that the second law of physics only applies to 'whole systems', not to heat transfer in each and every particular.

That's obscurantism, a practice that's gotten so common in science that anyone who states a matter plainly is now suspected of being a fake. That's a sad irony for it's been the academics, the pros, who trip all over themselves to explain and defend a theory that the evidence keeps contradicting.

So what has this left us with?

Just a sour, irrational attitude toward science that 'if it's incomprehensible, it must be true'.

---

[2] http://www.icecap.us/images/uploads/Lobert_on_CO2.pdf

If glass lets visible wavelengths of sunlight in but doesn't let invisible long-wavelengths (infrared) out, thus raising the temperature inside, then glass thermometers have been misleading us for centuries.

According to that same idea, glass thermometers necessarily register an extra 'greenhouse effect' and not the true temperature.

In reality however, no extra heating would come about even IF the glass were trapping infrared. The thermometer would simply take longer to adjust to changes of temperature, but it would NOT record a higher-than-actual temperature. As a Thermos demonstrates, trapping heat doesn't raise the temperature, it only sustains it.

The authors would much like to exchange ideas about the scientific basis upon which human-caused climate alarm is based, but sadly no debate—through no fault of the *Slaying the Sky Dragon* authors—has ever been entered into. Despite many detailed written exchanges, no scientific debate has ever been held between truly scientific skeptics and the obviously unscientific climate alarmists; only between the alarmists and the lukewarm skeptics, all of whom subscribe without question to the concepts of a 'greenhouse effect', 'greenhouse gases' and 'radiative forcing', as detailed below.

The Science

With no atmosphere at all, our moon is very hot in sunshine (over 100°C) and very cold in the shade (less than minus 150°C). The exact temps differ from zone to zone, but the ones given here illustrate the principle.

With earth receiving nearly the same amount of solar irradiation, our atmosphere acts as a cooling medium during the hours of sunshine and a warming blanket during the hours of darkness (alarmists love to abuse the blanket analogy—

using it to illustrate that the atmosphere is warmer during the day than it would be without one. But an actual blanket can at best maintain your body temperature, it cannot give you a fever. In net, it does not make you warmer, it just helps you retain your body heat...).

Global warming (which has by now—2010—been reversed to pre-alarm days), global cooling and all climate change is caused by the daily revolutions of our earth around its own axis, throughout which time the varying amounts of heat gained during the day and similar variations of heat lost during the night make the weather what it is: ranging from plus 50°C to minus 50°C (even more extreme in places), unpredictable beyond a few days (unless based on solar observations) and at times violent or totally quiet.

That's quite apart from the seasonal differences caused by the annual trip around the sun and the varying distance as our planet elliptically revolves around our sun—and we're not even considering even greater forces of influence.

Issue #1: By day, what heats a real greenhouse?

An actual greenhouse, whether made with glass or plastic sheeting, reaches higher temperatures inside than outside due to the restriction put on the internal air mass to disperse its acquired heat within the rest of the *open* atmosphere.

Even a wooden garden shed is warmer inside than the air outside. The internal air mass gains its heat from the total contents of the greenhouse, such as the soil or other ground cover material and all other objects within the space of the actual greenhouse. All matter within the confined space will absorb sunlight and cause the air within the confined space to warm up—initially by conduction, followed by convection.

The contents of the greenhouse thus gain their heat from direct sunlight, which is made up of a full spectrum of electro-magnetic radiation—including infrared.

Air is hardly warmed up by direct solar radiation (or any other radiation; radio, radar, TV, mobile phones, microwave ovens etc. etc. would otherwise not work) but is receptive to gaining or losing heat by means of conduction which in turn causes convection, carrying heat to ever greater heights— seldom the other way around.

Issue #2: What is a greenhouse gas?

The only true 'greenhouse gas' is air itself (oxygen and nitrogen). Gases such as water vapor and carbon dioxide have gained the reputation of being 'greenhouse gases' (GHGs) because they react (resonate) to radiation at various wavelengths and thus gain heat directly from sunlight as well as via conduction. In laboratory tests this means any enclosed air heats up more when there are more of these GHGs present in the space of the *enclosure* of the experiment. But there is no experiment possible that mimics the open atmosphere, by definition.

In the open atmosphere, quite the opposite of what we are led to believe, the so-called GHGs actually work to increase the scattering of any solar heat. Imagine two actual greenhouses—one with low humidity and the other with high humidity (any difference in level will prove the point). Actual experiments prove that a greenhouse with lower humidity takes less energy to heat. This is obvious as water vapor, a celebrated GHG, reacts to energy by warming up, but then dissipates this energy to the air that holds it. Opposite of what we're told, heat is not 'trapped'—it is dissipated. Carbon dioxide reacts in the same way as water vapor and dissipates any acquired energy. See below for further information about absorption.

Carbon dioxide is not a greenhouse gas; it does not absorb and store infrared or near-infrared in a way a sponge absorbs water and it does not transmit visible light—it is transparent to visible light.

Any energy that hits a carbon dioxide molecule will create, at the same instant, an equal and opposite emission spectrum, giving the casual observer the false illusion that energy has been 'absorbed', whereas it has merely been scattered.

Some of the energy that hits the carbon dioxide molecule may well increase the temperature of that molecule (depending on how the energy hits the alignment of the molecule), but that gained heat (this is theoretical only, it cannot be measured in our open atmosphere) will also be instantly dissipated by means of conduction with surrounding air molecules. At less than 400 parts in a million parts of air, those 400 carbon dioxide molecules would collectively need to reach several hundreds of degrees to warm the million parts of air by even a fraction of a degree, all at the same time, all over the world, all the time…all the while when warmer air rises and shares its gained heat with ever-higher-altitude molecules in our atmosphere.

The Pseudo Science

Apart from the climate change alarmists, many prominent skeptical scientists also make statements that are opposite to how the atmosphere works in reality, whilst some even make up new laws of physics to justify their incorrect assessments.

Clearing Carbon Dioxide of Blame

Herewith some quotes:

1. ...*all absorb heat radiation, and hence inhibit the cooling emission...*
2. ...*the earth is warmer than it would be in the absence of such gases.*
3. ...*adding to the 'blanket' that is inhibiting the emission of heat radiation...*
4. ...*This causes the temperature of the earth to increase until equilibrium with the sun is re-established.*
5. ...*the 2nd Law applies to the behavior of whole systems, not to every part within a system.*
6. ...*a photon being emitted by the cooler star doesn't stick its finger out to see how warm the surroundings are before it decides to leave.*
7. ...*The climate system is like the hot jar having an internal heating mechanism (the sun), but its ability to cool is reduced by its surroundings, which tend to insulate it.*
8. *In contrast, the infrared atmospheric greenhouse effect instead slows the rate at which the atmosphere cools radiatively, not convectively.*
9. ...*if there were only radiative heat transfer, the greenhouse effect would warm the Earth to about seventy-seven degrees centigrade rather than to fifteen degrees centigrade.*
10. ...*the sun shines on the top of the atmosphere, not the surface, and the emission of energy also comes from the top of the atmosphere, not the surface.*

The above junk science is refuted thus:

1. There is no physical mechanism by which a gas can absorb energy without at the same instant creating an

equal and opposite emission spectrum and in the open atmosphere of our planet there is in any case nowhere for energy to hide, other than in ice or water. Carbon dioxide cannot absorb and preserve energy. At no stage is cooling prevented and even if it was, that would not increase the originally achieved maximum temperature. A blanket can at best *maintain* your body temperature, it cannot add heat and give you a fever; it does not *make* you warmer, it just *keeps* you warmer.

2. Quite the opposite. The earth would be *warmer* if there were *no* water vapor in the atmosphere—and by some margin (but only during the hours of sunshine of course). Observational evidence can be seen on a daily basis when comparing maximum temperatures in deserts that have coastal fringes (e.g. Sahara, Namib and Atacama), where it will be seen that there is a direct link between humidity and maximum as well as minimum daily temperatures. *Absence* of water vapor allows more of the sun's radiation to reach the ground and thus create a *warmer* earth locally when compared to an atmosphere that holds greater water vapor and is at the same latitude. Conversely, the absence of water vapor will allow greater cooling at night whilst high humidity areas benefit from greater preservation of warmth, a sort-of 'greenhouse effect' in reverse.

3. That statement only holds true in high humidity areas and then only during the hours of darkness. The presence of water vapor creates a *cooler* daytime atmosphere and a less cold (*not warmer*) atmosphere at night. *At no stage is heat added nor created by the presence of water vapor or any other substance.* In any case, earth is already enveloped in the perfect 'blanket': the vacuum of space. Space is void

of matter and has no temperature of itself—we could not ask for a better insulation. As per #1 above, a blanket can at best *maintain* your body temperature, it cannot give you a fever and neither can a Thermos make its contents warmer.

4. If ever there was equilibrium between temperatures on earth and solar irradiance, the weather as we know it would cease. As is, solar radiation often varies more from mile to mile along any longitude or latitude than anyone could ever imagine and all climate-related 'averages' are purely mathematical entities that bear no relation to the actual situation at almost any point on our planet other than perhaps the coldest areas of the poles during their respective long periods of winter darkness when there is not enough energy entering the local climate system to create the greater variations witnessed in more temperate climate zones.

Just looking at the maximum and minimum temperature of a particular place in a moderate climate zone and deriving an 'average daily temperature' from such observations bears no resemblance to the ever-changing temperatures throughout the day. Between the observed maximum and minimum temperatures of the day, it could have hailed or snowed or rained or been overcast in several episodes. The struggle to reach equilibrium is what makes the weather so unpredictable —equilibrium can never be reached.

5. A brand new law of physics here, where parts within a system can behave contrary to the Second Law but the whole obeys. Only in 'climate science' can such chicanery be accepted as academic judgment. Thermal energy cannot flow into itself, only into something that has less

energy than itself. That's a law of nature, not a law of 'systems'.

6. A photon will not be able to raise the temperature of the object it is hitting if that object is already at an equal or higher energy level. In IPCC graphics, the photon that warms the earth and restarts the process is quite impossible (see the IPCC graphic in Figure 1). As per #5 above: thermal energy cannot flow into itself, only into something that has less energy than itself. That's a law of nature, not a moronic law of 'systems'.

7. Thermal insulation, in the setting of our open atmosphere, does not make the system average (or peak) one degree warmer than it would be without that insulation (the widely-accepted 'insulation' being 'greenhouse gases', not air itself (nitrogen, oxygen)). For a given energy input, a resultant maximum temperature is achieved and regardless of the amount or type of insulation, that maximum temperature cannot be increased. As per #1 above, a blanket can at best maintain your body temperature—it cannot give you a fever and a Thermos does not make its contents warmer, it merely slows down the rate of cooling.

8. An 'infrared greenhouse effect' (whatever next?) would need 'greenhouse gases' to store received radiation. Only water has that ability which is seen during the hours of darkness, not whilst the sun is adding energy, when water and water vapor soak up energy and prevent the atmosphere from warming as much as it would without water and water vapor (quite the opposite to what is being proposed).

9. A 77°C average surface temperature due to the purely radiative impact of the greenhouse effect? Radiant units do NOT combine in reality—101 W/m² directed at a blackbody that's radiating 100 W/m² raises its energy to 101 W/m², not 201—but in the much heralded Kiehl-Trenberth budget they DO combine. Let's look at the numbers, then.

According to the accepted Kiehl-Trenberth radiation budget (see Figure 2), the earth's surface averages 168 W/m² for solar absorption. K-T has the surface lose much of that energy by convection and evapotranspiration, though, so that 324 W/m² of back-radiated power bringing the surface up to 390 W/m², corresponding to 15°C. But in this case, we'll reduce convective and evapotranspirative heat loss to zero, which leaves us with the original 168 W/m². Now, within these parameters, how much extra back-radiation is required to bring the surface up to 77°C? SIX HUNDRED EIGHTY FOUR W/m², for a total of EIGHT HUNDRED FIFTY TWO W/m², which corresponds to 77°C.

(Bonus question: If the greenhouse effect generates enough radiative power to raise the earth's temperature to 77°C, but most of this heat is dissipated, then why is there no sign of this excess energy being blasted away from the earth? Satellites only see the earth emitting 240 W/m².)

The average solar irradiance for a blackbody earth—one that absorbs every photon the sun can provide—is 342 W/m², corresponding to an average temperature of 5.5°C. Here illustrious academia estimates conjure 852 W/m² out of nothing.

10. 'Radiative equilibrium' is an arbitrary construct to BEGIN with. You just subtract a planet's reflectance from the available irradiance and divide by four. **That's IT.**

There ARE no other steps. Since Earth reflects about 30% of sunlight, then, 1368 W/m² × 0.7 = 957.6 W/m². Dividing by four gives you 239.4 W/m², so that becomes earth's equilibrium figure and this corresponds to a temperature of 255 Kelvin. Now, is the earth's average SURFACE temperature 255K? No, it's warmer. So you say that 'somewhere up there' is where earth's radiative equilibrium is to be found, somewhere in the troposphere. That's silly, but once you convince yourself that the earth's temperature is NOT principally determined by the surface, you can convince yourself it IS determined by the atmosphere and that 'greenhouse gases' RAISE the 'equilibrium point' higher and higher. And as you see, you can even go as far as asserting that the surface absorbs no sunlight at all.

The Settled Science Unsettled

In spectroscopy, an absorption spectrum does not mean that energy is actually absorbed; it means an equal and opposite emission spectrum is created, indicating that intercepted energy is dissipated, scattered and re-radiated at different frequencies.

By looking only at the absorption spectrum gives the wrong impression, as so clearly illustrated by the overall emission spectrum of earth as seen by the satellites. Radiation input from our sun equals emitted radiation from the earth back into space, in expected accordance with basic and well-proven laws of physics.

Energy is not lost or created, whereas the widely and incorrectly accepted 'greenhouse' mechanism has it that carbon dioxide somehow re-radiates the same amount of infrared energy towards space as well as back to earth, thus apparently doubling the energy quantity—quite an impossibility—yet described in great detail by the greatest institutions on earth. See below for the latest list. The UN's IPCC graph of Figure 1 illustrates the classic and accepted view of the mechanism by which the earth gains heat, but this mechanism cannot exist; if it did, our energy problems would have been solved long ago by the engineering community.

"Surface gains more heat and infrared radiation is emitted again"—if only that were true!

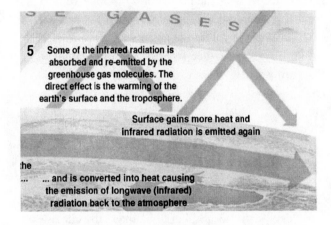

5 Some of the infrared radiation is absorbed and re-emitted by the greenhouse gas molecules. The direct effect is the warming of the earth's surface and the troposphere.

Surface gains more heat and infrared radiation is emitted again

the
... ... and is converted into heat causing the emission of longwave (infrared) radiation back to the atmosphere

Figure 1: IPCC Explanation for 'Greenhouse Gas' Heating.[3]

Whatever method of heat transfer is used, net energy flow will only take place if the receiver is cooler than the emitter, unless external energy is applied as is the case in refrigerators, for example. With earth emitting infrared energy and carbon

---

3

http://www.appinsys.com/GlobalWarming/GW_PART5_GREE NHOUSEGAS_ files/image001.jpg dioxide molecules re-emitting some of this energy back to earth, it is absolutely physically impossible for this re-radiated energy to warm the earth again. If that was not the case, the basic three laws of physics would need to be rewritten.

Yet this re-radiation of infrared is the very rock upon which the entire global warming panic rests. All who read this submission would do well to study the information on this page: First Principles of Heat Transfer[4]

The world has all too easily accepted greenhouse effect explanations that confuse the familiar reduction of CONVECTIVE heat loss with the production of radiative heat GAIN. A physical greenhouse merely slows down the normal cooling rate by limiting the volume of air in which heat loss is occurring. So here's a key feature to notice as the argument jumps to the atmospheric theory of a greenhouse effect: proponents will concede that the atmosphere provides no physical canopy—no actual pane of glass or blanket that confines heated air.

What's left, then?

Radiant energy itself. Rather than confining a fixed number of vibrating air molecules, the atmospheric 'blanket' the climate alarmists are arguing for is a RADIATIVE canopy under which infrared photons accumulate, and this extra energy buzzing around raises the temperature of all bodies under the canopy.

Thus the greenhouse effect amounts to a 'light battery' or generator continuously being fed by solar radiation, continuously being discharged at an EQUAL rate by terrestrial radiation, and yet it continuously AMPLIFIES the radiant energy inside it.

[4] http://jennifermarohasy.com/blog/2009/04/on-the-first-principles-of-heat-transfer-a-note-from-alan-siddons/

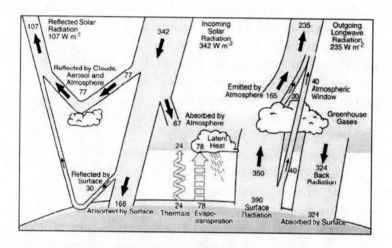

Figure 2: Kiehl-Trenberth Energy Balance[5]

As the Kiehl-Trenberth model shows: 235 units go in, 235 go out, and 324 are generated in between.

So the question naturally arises, "Is this even POSSIBLE?" Can photons of LIGHT be collected and multiplied like this? Can you turn on a flashlight, say, and put it inside a reflective Thermos, close the lid, and convince yourself that a million watts of radiative power will eventually be generated if you wait long enough? For that matter, has anyone ever INVENTED a device that captures light, like capturing wind in a bottle?

Or, do the laws of thermodynamics forbid this?
You decide.

We need to realize that blackbody equations are unable to predict a physical body's temperature to begin with; -18°C for the earth is a meaningless figure.

---

[5] http://www.cgd.ucar.edu/cas/papers/bams97/fig7.gif

No physical object radiates at a blackbody's rate, for one thing. And why? Because a real body has DEPTH: its response to light is not merely to heat up and immediately radiate the same amount in turn, but to conductively store the heat it acquires. Considering that the oceans alone are able to hold and circulate heat for decades, when do THEY reach a point of equilibrium with the radiation it has absorbed? Radiant energy budgets give it a year.

Who will get the message about the non-existence of an atmospheric greenhouse effect through to the 'climate change alarmists', the 'climate change skeptical' academics, the powers that be at EPA and most of the world's acknowledged institutions, NASA included, who all describe this non-existent 'greenhouse effect' with its 'greenhouse gases' in a language that mirrors the once celebrated justification for the existence of phlogiston and aether?

The Conclusion

There is not one single piece of evidence that supports the notion that carbon dioxide causes warming in the setting of our open atmosphere and in any case the physics involved in assessing a material's property will indicate that carbon dioxide, just like water vapor, is in fact a cooling agent ('fossil' fuel-fired power stations with their massive cooling towers are a classic illustration of the cooling power of water), an aid in the scattering of energy.

At least water vapor, with its thermal mass, has the ability to absorb energy and hang onto it (latent heat); carbon dioxide has no such ability.

In the reality of our open atmosphere it is thus the case that the only actual 'greenhouse gas' is air itself (mainly nitrogen and oxygen), whose presence allows an actual greenhouse to warm up. But quite opposite to an actual greenhouse, during the hours of sunshine it is this same air that keeps our open atmosphere cooler (compare the moon), whilst during the hours of darkness it prevents the atmosphere from cooling too rapidly (again, compare the moon). At no stage is our atmosphere warmer than it could possibly be due to the presence of water vapor, or carbon dioxide for that matter.

Trapped heat can never make the source of the heat hotter than it was in the first place—how could it?

The Near Total Deception

"Human-generated greenhouse gases are warming the earth but not as much as alarmists say" never was a good debating strategy for skeptical academics and it's probably too late for them now.

The only battle that remains is trying to limit the extent of emission controls on practical grounds, but the principle of emission controls has already been conceded.

Dissenters should have just stuck with the evidence: there is no sign of CO_2-caused warming at all and the *well-established physics* of the greenhouse gas theory must be confined to the dustbin.

Chapter 16

We Are Not Alone
by Hans Schreuder

IN THIS CHAPTER I'd like to present a number of short and pertinent quotations from eminent scientists from across the globe.

Let's first set the scene by going back one century to a noteworthy scientific event...

> *The astonishing discovery that atoms are mainly empty was made in 1909 at Manchester University by the indefatigable Ernest Rutherford. He had great courage as a scientist and was prepared to fly in the face of convention. Forced to explain the atom's mysterious emptiness, scientists had to jettison everything they had believed to be true for the previous two centuries. It was a seismic moment in the history of science.[1]*

---

[1] http://news.bbc.co.uk/1/hi/sci/tech/6914175.stm

Fast forwarding to 2009, Australian scientist Dr. Jennifer Marohasy states the following:

> Our understanding of the natural world does not progress through the straight forward accumulation of facts because most scientists tend to gravitate to the established popular consensus also known as the established paradigm. Thomas Kuhn describes the development of scientific paradigms as comprising three stages: prescience, normal science and revolutionary science when there is a crisis in the current consensus. When it comes to the science of climate change, we are probably already in the revolution state.[2]

From Dr. Nasif Nahle, USA:

> Throughout the last decade, supporters of the idea of an anthropogenic global warming or the impact of an anthropogenic "greenhouse" effect on climate have been insisting on an erroneous concept of the emission of energy from the atmosphere towards the surface. The global warming—greenhouse effect assumption states that half of the energy absorbed by atmospheric gases, especially carbon dioxide, is reemitted back towards the surface, heating it up.
>
> This solitary assumption is fallacious when considered in light of real natural processes.[3]

---

[2] http://jennifermarohasy.com/

[3] http://www.biocab.org/Heat_Stored_by_Atmospheric_Gases.html

That is, the longstanding paradigm says that because of trace gases like CO_2, the atmosphere heats the earth.

But this isn't true.

From Meteorologist William DiPuccio, USA:

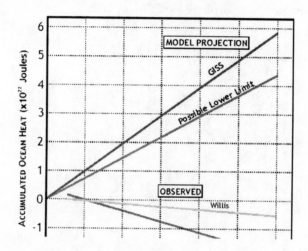

For any given area on the ocean's surface, the upper 2.6m of water has the same heat capacity as the entire atmosphere above it! Considering the enormous depth and global surface area of the ocean (70.5%), it is apparent that its heat capacity is greater than the atmosphere by many orders of magnitude.

The heat deficit shows that from 2003-2008 there was no positive radiative imbalance caused by anthropogenic forcing, despite increasing levels of CO_2. Indeed, the radiative imbalance was negative, meaning *the earth was losing slightly more energy than it absorbed.[4]*

There is no evidence of a recent global warming trend per se, despite increasing amounts of CO_2.

From Doctor of Meteorology Joe D'Aleo, USA:

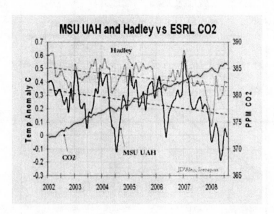

Given the current global cooling now in its 8th year, declining ocean heat content at least in its 5th year, sea level rises which have slowed or stopped, record rising Antarctic ice extent and rapidly recovering Arctic ice since the 2007 cycle minimum, a sun in a deep slumber, increasing evidence that CO_2 is a harmless gas that is in reality a beneficial plant fertilizer, you would think that this proposed legislation and ruling would in a sane world, have no chance of passing. But there is a huge political and NGO machine and all too compliant media

---

[4] http://climatesci.org/2009/05/05/have-changes-in-ocean-heat-falsified-the-gl obal-warming-hypothesis-a-guest-weblog-by-william-dipuccio/

and carbon crusaders like Al Gore and James Hansen and literally many billions of dollars behind making carbon evil and subsidizing unwise energy and carbon control solutions. [5]

A point that is reinforced by geologist Professor Ian Plimer, Australia:

The proof that CO_2 does not drive climate is shown by previous glaciations. The Ordovician-Silurian and Jurassic-Cretaceous glaciations occurred when the atmospheric CO_2 content was more than 4,000 ppmv and about 2,000 ppmv respectively. The Carboniferous-Permian glaciation had a CO_2 content of about 400 ppmv, at least 15 ppmv greater than the present figure. If the popular catastrophist view is accepted, there should have been a runaway greenhouse when CO_2 was more than 4,000 ppmv. Instead, there was glaciation. This has never been explained by those who argue that human additions of CO_2 will produce global warming.

The above makes a mockery of saying that today's level of atmospheric carbon dioxide is unprecedented.

From eWorldVu:

So, as American and European politicians prepare to fight global warming, Russia is preparing for a different world that may have much colder times ahead. If global

---

[5] http://icecap.us/index.php

temperatures continue to cool, it will be a cold war that Russia can win without ever firing a shot.[6]

From Russian News and Information Agency:

By the mid-21st century the planet will face another Little Ice Age, similar to the Maunder Minimum, because the amount of solar radiation hitting the Earth has been constantly decreasing since the 1990s and will reach its minimum approximately in 2041.[7]

From Geophysicist Norm Kalmanovitch, Canada:

It is inconceivable that even after a decade since global warming ended and seven years into a cooling trend with no end of cooling in sight, world leaders are unaware of these facts and are still pursuing initiatives to stop global warming. Something is terribly wrong with the official international science bodies such as the IPCC who have not come forward and properly informed the world leaders of current global temperatures. If in fact there is any validity to the claims of CO_2 increases causing warming; the fact that we are cooling at twice the rate that the climate models say we should be warming is a clear indication that natural forces are about three times stronger than the maximum possible effects from CO_2 increases.[8]

---

[6] http://www.eworldvu.com/international/2009/2/4/a-cold-war-that-russia-can-win .html
[7] http://en.rian.ru/science/20080122/97519953.html
[8] CCNet 78/2009

We Are Not Alone

From Professor Will Alexander, South Africa:

> *If there was strong evidence of undesirable changes, then the whole climate change issue would have been resolved long ago. The tragedy is that there is a worldwide policy in the opposite direction. Not only has the observation theory route been avoided, but climate change scientists and their organizations have adopted a policy of deliberately denigrating all those who practice it.*
>
> *[...] after 20 years of massive international effort (the overwhelming consensus), climate change scientists have still to produce solid, verifiable evidence of the consequences of human activities. They were unable to produce any scientifically believable, numerical evidence to support their theories. The periodicity in the data and the unequivocal solar linkage were not even addressed.*
>
> *This is not science. The whole climate change issue is about to fall apart. Heads will roll.*[9]

From Roy Clark, USA:

> *The 'radiative forcing constants' in the IPCC models are devoid of physical meaning. This approach is empirical pseudoscience that belongs to the realm of climate astrology. The results derived from climate simulations that use the radiative forcing approach may be of limited academic interest in assessing model performance. However, such results are computational science fiction that has no relationship to the reality of the Earth's*

---

[9]

http://anhonestclimatedebate.wordpress.com/2009/04/03/climate-change-%E2%80%93-the-clash-of-theories-by-Professor-will-alexander/

climate. Radiative forcing by CO_2 is, by definition a self-fulfilling prophecy, since the outcome is preordained with a total disregard of the basic laws of physics. An increase in CO_2 concentration must increase surface temperature. No other outcome is allowed and other possible climate effects are by definition excluded.

Based on the arguments presented here, a null hypothesis for CO_2 is proposed: It is impossible to show that changes in CO_2 concentration have caused any climate change to the Earth's climate, at least since the current composition of the atmosphere was set by ocean photosynthesis about one billion years ago.[10]

From John Ray (M.A.; Ph.D.), USA:

There is no such thing as a heat-trapping gas. A gas can become warmer by contact with something warmer but it cannot trap anything. Air is a gas. Try trapping something with it.[11]

From Geophysicist Norm Kalmanovitch, Canada:

There is not a single knowledgeable person in the world who cannot claim that CO_2 is beneficial to the environment. [...] There is not a single knowledgeable person in the world that cannot claim that the globe has been cooling since 2002. [...] There is not a single knowledgeable person in the world who cannot claim that with the past sea level rise of the last 8,000 years being

---

[10]

http://www.appinsys.com/GlobalWarming/EPA_Submission_R
Clark.pdf
[11] http://antigreen.blogspot.com/

less than four meters and based on the current rate of increase, the sea level rise by year 2100 will be in the order of just 16 cm (less than 7 inches). [...] Based on these three unequivocal facts it is clear that there is not a single knowledgeable person in government because governments refer to CO_2 as pollution and want to tax this "pollution" to stop the now non-existent global warming.[12]

These quotations are only a fraction of all the scientific work available to show that is an *indisputable fact* that there is *not one single observational item of evidence* to support the widely accepted idea that carbon dioxide *is* the cause of global warming or *even has an effect on climate change.*

Any and all evidence that has ever been presented to support the idea that carbon dioxide has an effect on global temperatures has been biased, opinionated and based on an agenda that pre-emptively dismissed alternative explanations.

Critically though, the global climate can neither be averaged nor can it be computerized and thus any and all scenarios coming from computer models are at best an exercise in computer programming but stand in no relation to reality, as clearly indicated by the totality of my submitted evidence.

Computer simulations regard the earth as a flat disk, without North or South Pole, without the Tropics, with few clouds and bathed in a twenty-four-hour haze of sunshine. The reality is two icy poles and a tropical equatorial zone, with each and every square meter of our earth receiving an ever-varying and different amount of energy from the sun, season-to-season and day-to-day. This reality is too difficult to input to a computer.

Did you realize that?

---

[12] CCNet 68/2009

If carbon dioxide really is such a danger to mankind, as the U.S. EPA would have us believe, then the upcoming 2012 Olympic Games should be cancelled, as well as all other big sporting events, as well as all road transport and all air transport and all coal- and gas-fired power stations should be shut down, all boats and trains to be halted—in fact, we might as well stop breathing too.

Clearly there is no need for such drastic action and clearly atmospheric carbon dioxide at even 400 PPMV is not dangerous at all, why, when we breathe out the level is a whopping 50,000 PPMV.

From the word go, the UN IPCC has provided us with scenarios based on the principle of perpetuum mobile by clearly indicating that the earth is getting warmer due to re-radiated infrared energy from the increased levels of carbon dioxide.

That scenario cannot physically exist.

The sun provides the energy to warm the earth and the only possible effect that carbon dioxide *could* have on the atmosphere is to increase heat *dispersion* and thus cause cooling. But at 400 PPMV the effect would not be measurable.

As a further rebuttal of the influence of carbon dioxide over the climate, the alleged IPCC greenhouse effect is a non-existent effect. No greenhouse, whether made from glass, plastic, cardboard or steel will reach a higher inside temperature due to the magic of re-radiated infrared energy. If it did, engineers would have long ago been able to design power stations made from air, mirrors and glass, extracting more energy out of it than was put into it—*if only!*

In conclusion, then, a century after Rutherford's momentous lecture, I urge the reader to consider nothing but the facts before them. Those facts are that carbon dioxide does not and cannot cause global warming, the currently accepted paradigm notwithstanding.

Any and all schemes to reduce carbon dioxide emissions are futile in terms of having an effect on reducing global temperatures or affecting the climate and any and all carbon trading exchanges are fraudulent exercises amounting to no more than hidden taxation.

One other relevant scare: Ocean acidification

Besides the alarm over the climate, there are alarmist screams over ocean acidification due to increased levels of atmospheric carbon dioxide. It is actually the carbon-dioxide-rich oceans that drive atmospheric carbon dioxide levels, so a quick word on this matter. The companion volume has full details on this and other scare stories.

Erl Happ makes this point about the ocean power game:

Tropical sea surface temperatures respond to the change in surface pressure across the globe and in particular to the differential between mid latitudes and the near equatorial zone. The southern hemisphere and high latitudes in particular experience marked flux in surface pressure. This leads directly to a change in the trade winds and tropical sea surface temperature.

Is there evidence that the activity of man (adding CO_2 to the atmosphere) is tending to produce more severe El Nino events? The answer is no. The flux in surface pressure is responsible for ENSO and for the swing from El Nino to La Nina dominance. In spite the activities of man, the globe is currently entering a La Nina cooling cycle testifying to the strength of natural cycles and the relative unimportance atmospheric composition in *determining the issue (if the much touted greenhouse effect exists at all).*

Is there evidence that the ENSO phenomenon is in fact 'climate change in action', driven by factors other than the increase of atmospheric CO_2? Yes, it appears that whatever drives the flux in surface atmospheric pressure drives ENSO and with it, climate change.

Is recent 'Climate Change' driven by greenhouse gas activity? No, it appears that the cause of recent warming and cooling relates to long-term swings in atmospheric pressure that changes the pressure relations between mid and low latitudes thereby affecting the trade winds that in turn determine the temperature of the Earth's solar array, its tropical ocean, and ultimately the globe as a whole.[13]

At the end of this chapter I hope you realize the power the sun and the oceans have over our climate and that part of the atmosphere in which we all live and breathe.

The oceans have moods influenced by our moon and sun, causing tides, wind, storms, calm, heat and cold. The as yet uncounted thousands of 'black smokers', underwater volcanoes and assorted other cracks in our earth's underwater crust all have their influence.

To even think that humans can influence the vastness of the oceans is to give ourselves a level of influence we simply do not possess. What we do possess is the power to cause localized and often serious water pollution, with local loss of aquatic wild life or in the case of oil spills, devastating results for sea birds as well.

Those pollution instances are usually caused by industrial accidents, with the recent oil spill in the Gulf of Mexico a classic example. As we now know though, the actual damage

---

[13] http://icecap.us/index.php

there was far less than the anticipated catastrophe, as is so often the case with environmental 'problems'. It's all too easy to exaggerate and scaremonger.

Same with climate change: there is no cause for alarm over a short-term minute increase in global temperatures (which in any case has by now—2010—turned into a global cooling trend, which is much more alarming for all of us) and the human race can do nothing to cause either warming or cooling—Mother Nature will do her own thing, as she always has.

As Derek Alker so aptly observed:

The K&T earth energy budget illustrates a dead planet, as do all similar budgets. None of them take the energy absorbed by life itself into account.

Chapter 17

Climate and the Geo-Nuclear Connection
by Joe Olson

TO BEGIN ANALYZING the possible climate change factors, it is necessary to recognize that weather, and the compilation of weather, referred to as climate, is just an observation of the final end reaction of a large number of interactions by all of the primary forces of the Universe.

It is hubris, stupidity, or intentional deception to ascribe human actions and one atmospheric trace molecule with control of this vast interacting system.

The fundamental Universal forces are gravity, magnetism, electro-magnetic radiation and nuclear attraction. All play a role in this final ripple on the pond of reality—called climate.

To understand this complex interaction has required analysis of a wide range of data, some from disparate and suppressed sources. All of this material is fact and logic-based, even if not common knowledge.

This discussion will be limited to minimum descriptions necessary with references for continued study and projections on the immediate future of this new, comprehensive Earth Science Theory.

239

To advance the hypothesis that human-produced atmospheric carbon dioxide was the primary climate-forcing system required the fabrication of two false sets of data. One was a thousand year 'guess' of CO_2 levels and the other, a matching thousand-year computer-generated temperature graph. With the cause-and-effect link established, it was only necessary to violate a number of established principles of physics for the climate charlatans to prove their case to the unwitting public, politicians and media pundits.

To unravel the climate puzzle we will first describe the three fundamental forces, then their individual interactions affecting climate and finally, how all of the forces combine to form the harmonic balancing act which nurtures our biosphere.

Before dismissing this line of thought as over-arching, consider the readily apparent effect of these three forces on climate.

Climate is, in the final analysis, a heat flow equation effected by fundamental forces. Gravity holds the Earth in an elliptical orbit around the Sun accounting for a 10% variation in solar radiation due to this variation in distance. Consult any chart on the eccentricity, obliquity and precession of Earth's orbit and it is apparent that gravity is a force in the climate equation, but this force is not the unnoticed factor in this discussion.

The Earth's magnetic field shields the planet from solar radiation and is also subject to variations. Nuclear fusion and fission reactions heat the Sun and this planet's internal fission reactions also heat the Earth. All of these nuclear reactions are variable.

It is the combination of these factors that create our climate, when combined with one additional factor. Life itself exerts a powerful factor in the climate equation far in excess of the minute fraction that human-produced gases can exert.

The processes of photosynthesis, transpiration, respiration and organic decay have measurable climatic effects, but have been reviewed by others and are not the subject of this chapter. In any complex system there are primary drivers with positive and negative feedback and buffering systems. We will introduce these elements, but can in no way establish all of the still largely unquantified relationships.

We will begin the discussion of fundamental forces with the suppressed origin of our solar system first proposed by Dr. Oliver Manuel, known as the Supernova Origin Theory[1].

Four billion years ago, an iron-rich supernova exploded, sending planet-forming material out in an equatorial belt which would soon form our solar system ecliptic plane and eventually our planets. Gravity and magnetism held higher concentrations of iron and other heavy elements closest to the largest fragment of that exploded supernova, which would reform to become our Sun. The core of Earth is not an amorphous blob of molten iron.

Earth's core is a single, cubic crystal of iron molecules which form a giant permanent magnet. Each atom of iron is a dipolar, miniature magnet, which in a liquid state, aligns with the dominant existing magnetic field. Atoms have a natural nuclear repulsion which increases with temperature and the level of this kinetic molecular energy is displayed in the three states of matter as solid, liquid and gas.

As each molten liquid iron atom joins this cubic crystal lattice it aligns based on the already established crystal magnetic field. This iron lattice allows the maximum density in both weight and magnetism. Proof of the Earth's crystal

---

[1] The Sun's Origin, Composition and Source of Energy, O. K. Manuel, *Abstract #1041, 32nd Lunar & Planetary Science Conference,* Houston, TX, March 12-16, 2001

lattice core was first mentioned by Song and Sun[2] and it is the extension of accepted science that this would form a natural magnet and that similar crystal magnetic cores existing in the Sun and inner planets as fragments of a previous crystal core in the supernova of our solar system origins.

The six inner planets all have iron cores and varying magnetic fields. A spinning magnet produces an electrical current and an electric current itself produces a magnetic field. Mercury, Venus and Mars have all lost their molten mantles and their crystal cores are now locked into the planet surface rotation causing reduced and constant magnetic fields. The Earth's permanent magnet crystal core is floating in a molten rock globe with a thin solidified crust.

There are variations in the spin axis and spin rate which are manifested in variations in the magnetic north location and the total magnetic field strength. The Earth's core is actually spinning 3 degrees per year FASTER than the planet[3]. A 1500 Km (950 mile) magnet need not spin very fast to create a significant electric and magnetic field. The spin rate should also have a predicted influence on the internal fission rate. Since all of the primary forces are interactive it is necessary to develop the theory linking these forces.

The Sun, Jupiter and Saturn all have crystal iron cores floating in a gaseous sphere with an even greater degree of freedom of movement. Gravity, magnetism and electro-magnetic radiation all vary as a square of the distance and are thus non-linear variations. Due to distance and field strength,

[2] Researchers Confirm Discovery of Earth's Inner, Innermost Core, *Illinois News Bureau,* James E. Kloeppel, reference to Song & Sun, News.illinoi.edu/news/08/0310core
[3] Earth's Core Rotates Faster than Surface, Ker Than, *Live Science,* 25 Aug2005, TechMediaNetwork.com

it is Earth and Jupiter that exert the most influence on the Sun's magnetic core. The Earth-Jupiter conjunction occurs every 11.5 years. Every 11.5 years the Sun experiences a magnetic pole reversal. Every solar reversal creates measurable changes in solar radiation and the solar magnetic field. A complex and exhaustive analysis of the relationship of our solar system movements and the resulting periodicity is given in *Empirical Evidence for a Celestial Origin of the Climate Oscillations* by Nicola Scafetta[4].

This work research establishes fundamental 11, 22 and 60 year cycles which are reflected in temperature graphs, but not reflected in CO_2 concentrations or IPCC predictions. Portions of the climate geologic record do appear to have 'astronomical' precision, but there are anomalies. We will accept these celestial factors within the previous 500,000 year Milankovic Cycle and cosmic ray variations in the previous 600 million year Phanerozoic Period.

The question raised, but unanswered, is the method that cosmic rays can use to have such measurable effects on Earth's climate. For that, we will explore the next level of primary Universal forces and their collective effect on our climate.

Toy globes sat on desktops world-wide for centuries with the puzzle like fit of eastern and western Atlantic shorelines begging the question, are the continents 'drifting' from a former merged land mass ? This question was first asked by Flemish cartographer, Abraham Ortelius in 1596.

Scientific 'consensus' on drifting continents took 370 years of discovery and debate. The motive force in plate tectonics was not fully understood until the second generation of GPS satellites was in orbit. The first generation of GPS

---

[4] Scafetta, N., Empirical Evidence for a Celestial Origin of the Climate Oscillations and its Implications, *Journal of Atmospheric and Solar-Terrestrial Physics (2010)*, Doi:10.1016/j.jastp.2010.04.015

offered place locating within 0.5 meter (18 inches) and offered the exciting possibility for land surveyors of eliminating cumbersome two-station back-sight for horizontal control and lengthy level loops for vertical control. In short, it was hoped that a single satellite reading would completely locate any point on the planet. The second generation did provide accurate horizontal control, but the vertical variation continued.

It was not until this satellite based reference system was in place that any scientist realized that the thin crust of the Earth was lifted daily by the Moon's gravity, much like the oceans—this effect was called 'Earth tide'. This lifting action pulls the plates to the west, opening the ocean floor rifts which then quickly fill with molten rock from below. This molten rock instantly solidifies and creates the ratchet effect on the opposite side of the plate along a subduction zone. This daily crust movement sends pressure waves thru the molten rock and pumping action to stir the mantle and release trapped air pockets.

We now have a stirring force of the spinning, crystal iron core and a pumping action from above, working on this plastic molten rock layer. Any force that is capable of 'floating' entire continents must be considered in any proper 'heat flow' analysis.

We are standing on a planet with 1.08 quadrillion cubic Km (259 trillion cubic miles) of molten rock ranging in temperature from 1400°C (2500°F) near the surface to 6000°C (10,000°F) at the core. Our planet has 1.29 billion cubic Km (310 million cubic miles) of ocean. Volume measurements are not useful for the compressible gas atmosphere, but the total atmosphere is 0.00008% of the total planet mass. Carbon dioxide is 0.04% of that portion of gas and human CO_2 is at most only 4% of that insignificant amount. In terms of possible heat storage, the CO_2 storage capacity is not even measurable. Carbon dioxide can absorb infrared sunlight only within two narrow spectrum bands[5]. One of these bands is shared with water vapor, so this amount of radiation would be absorbed in the atmosphere regardless of CO_2 content unless there was zero humidity.

Any radiation absorbed on its way to the Earth, never can warm the surface and can never be re-radiated at night. The wavelengths of energy that are radiated back into to space at night have a very short dwell time in the atmosphere on its way, in accordance with the Second Law of Thermodynamics from a state of higher energy (Earth) to lower energy (outer space).

There is a finite amount of IR radiation available to absorb or re-radiate, so there is no linear relationship. Doubling the CO_2 does not double this limited heat retention.

We have now established the parameters surrounding the most glaring error in all of the climate change models, the complete absence of Geo-nuclear produced energy.

The Universe is a nuclear waste ground. Nuclear decay is all around us. Our planet is constantly bombarded by solar and galactic particles from without and fission released particles from within.

The Universe seems in a mad dash to become a uniform distribution of Hydrogen atoms. The 259 trillion cubic miles of 'solid' material that forms our planet is 4 parts per million of Uranium, or 2.9 million cubic Km (700,000 cubic miles) of nuclear fuel. During fission each Uranium atom has 92 proton-neutron pairs which during fission could form 92 Hydrogen atoms, or 46 Helium atoms or 16 Carbon atoms or

---

[5] Carbon Heat Trapping; Merely a Bit Player in Global Warming, R. J. Petschauer, IEEE Senior Member,
http://vipclubmn.org/Documents/GlobalWarmingArticle.pdf

12 Oxygen atoms, along with highly charged particles capable of busting apart an adjoining atom or two. Each Uranium fission reaction produces two-million times the energy of a Trinitrotoluene (TNT) molecule, hence explosive power measured in megatons of TNT.

The fission breakdowns occur in stages, with as many as a dozen daughter reactions. Some of the products of these reactions are very stable Inert (Periodic Table Group VIII, or Noble) gases which are non-reactive.

In nature, these inert elements do not form chemical bonds with other atoms to form molecules. The very presence of these gases is proof of Earth's fission forces.

Radon is one of these inert fission by-products and unstable, with a half-life of only 3.8 days. If you had a kilogram (2.2 lbs) of Radon it would disintegrate to just a 7.8 grams (0.27 oz) in just 21 days. Given travel time from creation to release at the surface, there must be an enormous production of Radon for there to be any detectable levels at the surface. Helium is the lightest element when released in the atmosphere and is 1/6th the weight of air. Small quantities of Helium exist in the tops of underground natural gas reserves, but all Helium is in a desperate race to outer space.

The complete array of fission-produced atoms is referred to as 'elemental' atoms and the compounds that they form in the Earth's cauldron are referred to as 'elemental' compounds. We will return to the compounds in a moment, but for now the important fission-climate link.

All of the heavier elements are subject to nuclear bombardment and breakdown. The rate that this occurs in a laboratory is termed the half-life, but is not constant in nature. Increases in the rate of particle bombardment reduce the half-life. This occurs at a controlled rate in a nuclear reactor and at near spontaneous rate in the 'chain reactions' of a nuclear bomb. The heavy elements in the Earth's core are subject to constantly varying levels of particle bombardment from solar, galactic and Earth fission produced particles.

The Earth's fission energy is substantial and variable. This internal energy is not included and any climate model energy flow.

It is an error in the energy balance equation if it is assumed as zero. It is an error in the climate change equation if it is assumed constant. There has been an order of magnitude increase in the estimated amount of this energy since the first IPCC models were developed[6][7]. There has been no IPCC correction based on this new information.

This fission energy is displayed on every continent with hot springs and geysers, thru ocean floor vents and thru volcanoes. Despite claims that certain of the manifestations are 'Old Faithful' level of reliance, they are all highly variable.

This is a reflection of the variation at the source. The Uranium and other heavy element atoms that are stirred and pulsed in the molten mantle are not uniformly distributed or activated. It is therefore possible that regional hot spots can develop raising ground temperatures slightly and altering air patterns.

This produces regional weather anomalies. Similar changes at the sea floor alter ocean currents. These forces are then coupled with solar events to produce the short term weather events.

---

[6] *First Measurements of Earth's Core Radioactivity*, Celeste Biever, New Scientist, July27, 2005,
newscientist.com/article/mg18725103.700
[7] *Potassium-40 Heats Up Earth's Core*, Belle Dume, Physics World, 07May2003,
Physicsworld.com/cws/article/news/17436

Dramatic changes in these fission reactions are the cause of long term climate events like Ice Ages. If you refer to the Geocraft.com 600 million year chart below, you will first notice there is no apparent correlation or causation between CO_2 and temperature. You will also notice two periods of very dramatic temperature drops at 450 million and 300 million years ago.

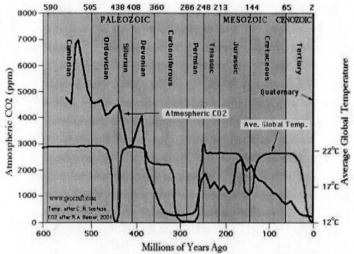

Figure 1: Millions of Years of CO_2 vs. Global Temperatures[8].

These are the 'Popsicle Planet' eras when ice covered all but the 30 degree equatorial band. Polar ice caps extended south to Miami and north to Rio de Janeiro.

Now moving to the 500,000 year Milankovitch Cycle chart you can see dramatic change forming three distinct

---

[8]

http://www.geocraft.com/WVFossils/Carboniferous_climate.html

100,000 year-long glacial periods, separated by 12,000 year long inter-glacial periods.

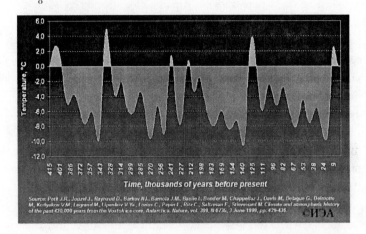

Figure 2: 400,000 Years of Temperature Data from Ice Cores[9]

The atmosphere masks some solar radiation variations and mankind lacked the instrumentation for accurate solar measurements until launch of the NASA's Helios and Solar Heliograph Observatory (SOHO) satellites. During their more than thirty-year measurement period, these satellites have measured only a 0.1% variation in TOTAL solar output; however there is substantial change within certain wavelengths and in solar particle flux.

It is inconceivable that there have been massive fluctuations in total solar output that caused these previous temperature change events. It is also undeniable that during the thirty years of near constant solar radiation as per SOHO

---

[9] http://www.iccfglobal.org/ppt/Illarionov-01-10-04.ppt, Slide #43

measurements, there have been three solar magnetic reversals and dramatic El Nino and La Nina events that effected Earth's climate. All of these planetary thermal changes occurred with constantly rising CO_2 levels. The most plausible explanation is then that some solar or galactic particle bombardment change altered the rate of Earth's internal fission heater. It is also undeniable that humans have played no part in any of these previous climate cycles.

Realizing that to accuse the 10 to 20 parts per million of CO_2 being the sole driving force for climate change would be UNBELIEVABLE, the IPCC cabal then added another atmospheric player, water vapor. Never more than 5% of the total atmospheric gases, water vapor and its effects are readily observable. We have all noticed that there is less heat loss on winter nights with overcast skies. The IPCC scientists insist this is proof of greenhouse gas behavior. Better trained scientists credit the denser, water vapor laden air, with greater thermal mass, which slows heat transfer. In discussing the invaluable role of water vapor to climate, we must review some basic water properties.

Rare among compounds, is the fact that water exists in all three phases in our natural environment. To change state from solid to liquid to vapor requires heat, or energy input. One gram of solid 'ice' at 0°C requires 80 calories to become 0°C water. It then takes 100 calories to raise that near frozen water to near boiling water at 100°C. An additional 540 calories will raise that 100°C very hot water to 100°C very cool steam or vapor.

The heat required to change state is termed 'latent heat' and individual changes are latent heat of liquefaction or latent heat of vaporization when adding energy.

The reverse processes are termed the latent heat of condensation and latent heat of solidification. These same thermal parameters apply to water in our natural environment. When water is evaporated from cooling towers, evaporative coolers or even our own body perspiration, it removes 2257 kilo-Joules/Kg (890 BTU/lb) of heat per unit of water. As water sublimes directly from ice to vapor, or is evaporated from surface water it is removing heat from the Earth's surface. When it condenses into droplets in clouds or solidifies to snow, sleet or hail in clouds, it is releasing that energy high in the atmosphere.

With this enormous heat transfer all adding to the planets cooling, it is deceitful to claim this vapor is warming the planet. Heat that is released high in the atmosphere DOES NOT then radiate back to warm the planet. Thermodynamic Laws demand that heat to flow to outer space. Even if you were to consider heat flow to be random, there are two directions of travel along three different axes. Just random probability would dictate that only one out of six possible directions for this heat would be back toward Earth.

Suppose that climate science did want to estimate the true flow of energy on our planet. We have a well-established rate of solar input and two possible ways of measuring water vapor heat flow. Within the accuracy range allowed, it is safe to assume a constant amount of water. Water is highly reactive, readily breaking the Hydrogen-Oxygen bonds to form other compounds, reducing the amount of water. Water is subject to nuclear particle breakdown and some loss in the upper atmosphere to solar wind.

There is also the addition of elemental water from fission-produced atoms. Neglecting those minor variations, we could estimate the evaporation-sublimation side of the equation or the condensation-solidification side of equation to determine the water vapor heat flow. Given the tons of rain, snow, sleet and hail that fall on the Earth every second it is obvious that atmospheric science has yet to grasp the full extent of Earth's true energy flow.

251

When the IPCC team chose to portray the last 1,000 years of temperature and CO_2 as two matching hockey stick graphs—it was a willful attempt to deceive. The hockey stick maker chose what has been referred to the 'Most Important Trees on the Planet' to represent past temperature[10].

Only these few trees, of the hundreds of tree rings examined provided the 'desired' temperature result. That result was the elimination of the Minoan, Roman and Medieval warming periods. Roman records of vineyards, Greenland dairy barns under feet of present day snow and Japanese cherry blossom records all refute the seven selected tree ring hypothesis.

Many scientists chose to believe what our ancestors have left as proof. The same skeptic view applies to ALL paleo-climate data. All predictions of past conditions are proxy data and subject to two obvious error paths. First is the level of degradation of the samples over time and second that past conditions were very similar to current conditions. Either of these conditions is a fatal error in correct science, yet these warmist advocates violate both. We have previously mentioned the constant nuclear decay that surrounds us. Physical erosion is another partner in planetary decay.

The upper atmosphere is subject to constant erosion from solar wind and Nitrogen ionization by cosmic rays. When the Earth has its periodic magnet declines and reversals, there is even greater erosion.

When major asteroid impacts occur, waves of atmosphere are pushed beyond the force of gravity and are lost to space. When the Yellowstone super volcano exploded 640,000 years ago it blew one thousand cubic Kilometers (240 cubic miles)

---

[10] NAS Panel #2: Bristlecones, Steve McIntyre, climateaudit.org/2006/06/29

of rock into space. By the Venturi effect, massive amounts of atmosphere would have been drawn into space with that rock.

The newly discovered polar ice caps on the Moon are just eons of accretion of full Moon orbits thru Earth's solar wind vapor trail. There is largely ignored evidence that the past atmosphere was vastly different than today's by comparing winged flight. Successful flight involves four components, lift, thrust, drag and gravity. It is safe to assume that gravity was not significantly different in the past. It is also safe to assume that nature did not evolve from lighter bones and stronger muscles to reduced levels of both today.

With the other three variables constant, there is only one option for the fossil record. Compare the 65 million year old *Pterodactyal, Quezalcoatus Northropi* and its eleven meter (hirty-six feet) wingspan with today's largest Peruvian Condor wingspan of five meters (sixteen feet). In addition compare the *Meganeura Dragonfly* with a half meter (twenty-inch) wingspan and today's Atlas Moth with its quarter meter (eleven-inch) wingspan. The obvious conclusion is that there was TWICE the atmosphere in the Triassic Era, which provided additional lift.

Visit the Geocraft chart again and verify that temperatures were also significantly warmer during the Mesozoic Period with significantly higher levels of CO_2. With ferns the size of houses and lizards the size of busses it is hard to argue that either warmth or CO_2 increases are harmful to the planet or to life. This Plate also shows the irrefutable proof of a far colder planet at the Ordovician-Silurian and the Carboniferous-Permian boundaries.

Far more than simple cosmic ray changes are required to explain these 'Snowball Earth' conditions. Nuclear fission reactions are self-sustaining within limited conditions. The break-down of heavy elements gives off high speed particles that can create an adjoining atom's decay. In a melt-down

253

condition there is enough energy produced to explode and send a Yellowstone mountaintop into space.

When particle bombardments from internal and external sources drop, the Earth can scarcely warm the equator with just solar energy.

During Earth's Snowball Planet phases, solid ice may have extended to the ocean floor, limiting all life to just the equatorial green belt of unfrozen ocean and snowpack-free land masses. There had to have been a significant change in energy for the planet to then transform to the lush tropical paradise of the Triassic Period. Any understanding of climate must be comprehensive.

There is absolutely no correlation or causation with any of these past events with carbon dioxide levels and most certainly NO human forced cause of change. To embrace the IPCC and AGW theories of climate you must first accept that human's powers beyond the fundamental forces that we now know form the complex interplay of climate.

The overt suppression of debate on this subject by government funded 'big' science and commodity market driven 'big' media has forced independent minded scientists world-wide to re-examine a wide range of science.

This has been a journey of understanding for all humanity. Ideas which are presented in a more formal setting in this book were first presented and evolved as articles posted on the internet. As typical with all good science, discovery prompted new insights. After a lifetime of science study and thousands of hours dedicated to just this issue, there is one previous statement that I will now revise.

In the article, written over a year ago, titled *Humanity's Last Chance for a Fairy Green Future* there was mention of the analogy of our celestial symphony being like a Junior High Band Concert. A comparison was made that CO_2 was a third chair violin, making the only sound that the proud IPCC parent could hear. A more proper analogy would be that CO_2 is the dust in the concert hall. This dust can be raised by the music, even visible under the correct lighting conditions. This dust could be discovered by scientists years later to have had 'movement' during the concert.

This dust did not create or alter in any significant way the true sights and sounds of this concert. Carbon dioxide is in the end the most basic form of dust and the cornerstone of all carbon based life forms.

To demonize such an innocent and vital component of life betrays a deeper seated hatred for life. This is the unrepressed primordial instinct to dominate and control the uneducated masses. That urge is reprehensible enough, but to twist science into a weapon to further that end is a true crime against humanity. In the end, climate is very much like a symphony with dozens of forces making up the various players.

Under the IPCC maestro, whole sections of this orchestra are silent. The IPCC is a false prophet peddling Faux Science. We must demand that all future discussions of climate science include the fundamental forces just described.

It would not be enough to just examine the flawed science of this man-made problem. Equally troubling is the science truth behind the proposed 'green energy' solutions and the revelations of the truth about 'carbon' energy. We will begin with an analysis of the Faux Science of solar cells and batteries. We will then return to those 'elemental' compounds brewing in the Earths chemical refinery. These are the feedstock to Earth's greatest gift to humanity, Abiogenic Oil.

Additional Defects of the Green Machine

There can be little doubt that the entire human caused climate change issue has been an intentional politically motivated fraudulent movement. This has evolved from an elaborate network of direct government involvement and indirect government funding to provide the illusion of 'consensus' that would be beyond any further debate.

Fortunately for the future of science, truth and humanity, many honest scientists and analysts from many lands have objectively looked at the hypothesis of Anthropogenic Global Warming and found it invalid.

No analysis of this failed hypothesis is complete without examining what has been endorsed by the AGW supporters as the 'solution' to this non-existent problem. The proposed 'Green Energy Solution' is as defective as the AGW science, with defects so obvious that endorsement must also qualify as intentional deception.

The 'Green Energy Solutions' are primarily focused on wind energy and solar energy, with fictional claims for future tidal and geothermal which have been fictional claims for over a half century already. If stopping the release of hydrocarbons was the highest priority, then nuclear and hydro-electric would be considered 'green', but these power sources have long been on the eco-zealot 'hit' list.

The Eco-religion could not allow reclassification of these two carbon-free energy systems and maintain peace between the devout tree-huggers and the obsessed warmists. Demise of the warmist orthodoxy will reopen debate and action on all reliable systems of energy production. For now, some limited further analysis of the unreliable 'green energy' systems, with full analysis in future books by our team.

Climate and the Geo-Nuclear Connection

Volumes have been written on the defects of wind-driven electrical generation. The wind energy standard is a net output of less that 25% of rated capacity with constant fluctuations in power. Add in bird and bat strikes, noise and visual pollution, transmission losses and foundation failures and you have just a portion of the wind energy flaws.

Every existing 'wind farm' is its own testament to failure. The truly functional wind farm is an illusion that humanity cannot afford to waste our resources on.

As mentioned there are volumes of material on wind energy available, but what is not as well understood are the defects of solar energy. As alluring as the premise may be, the promise of solar energy is not free.

The first solar cell was created in 1883 by Charles Fritts using a sheet of Selenium with thin Gold facings. The Sun radiates approximately 1,000 watts per square meter at maximum. The Fritts cell produced 10 watts per square meter or 1% efficiency. The Russell Ohl patent of 1946 is considered the first modern solar cell.

Today's solar panels are high purity Silicon with a light doping of Phosphorus and Boron to provide breaks in the Silicon for electron movement. Silicon crystal is highly reflective, so the solar facing side must be treated with an anti-reflective coating. In addition, the solar array includes a conducting surface grid covered with glass for protection against the dust, droppings, weather and other environmental hazards.

All of these conditions limit some of the incoming light. Only certain segments of the solar spectrum activate the flow of electrons and the net result is 10% efficiency, or approximately 100 watts per square meter. Efficiencies as high as 40% are available with exotic materials, but then one must address the 'high cost of free', which applies to every 'green' technology.

Silicon, Phosphorus and Boron are common elements, but to mine, refine and bring on line has a cost. That cost is reflected in 'cost payback' of five to seven years depending on the system. The total system life is twenty years. But these costs are based on low cost carbon based energy systems providing these materials.

Much like paying your Visa bill with your Master Card, this parasitic 'clean' energy cannot provide the 'spare' energy to avoid 'dirty' energy. There is a certain loss of electrons in this system and power production erodes over time—after twenty years they are useless. Sunlight is not converted to electricity. Sunlight erodes molecularly-stored potential energy from the embedded Phosphorus atoms until there are no spare electrons left.

The search for scientific truth in one field often leads to unexpected insights into other fields. In researching the ignored or vastly underrated role of Earth's nuclear fission in climate change another truth became self evident.

Matter is neither created nor destroyed.

In future publications, I will point out that there can be no doubt about the fact that what is routinely referred to as 'fossil fuel' is incorrect and even oil is a renewable resource.

Chapter 18

Climate Thermodynamics
by Claes Johnson[1]

Global Climate by Navier-Stokes Equations

Thermodynamics is a funny subject. The first time you go through it, you don't understand it at all. The second time you go through it, you think you understand it, except for one or two small points. The third time you go through it, you know you don't understand it, but by that time you are so used to it, it doesn't bother you any more.
—Physicist Arnold Sommerfeld (1868-1951)

GLOBAL CLIMATE IS the end result of a thermodynamic interaction between the atmosphere and the ocean—with radiative forcing from the Sun, gravitational forcing from the Earth (and the Moon) and dynamic Coriolis forcing from the rotation of the Earth.

The thermodynamics is described by the *Navier-Stokes Equations* (NSE) of fluid dynamics, for a variable density

---

[1] Computer Science and Communication, KTH, SE-10044 Stockholm, Sweden.

incompressible ocean and compressible atmosphere, expressing conservation of mass, momentum and energy.

The atmosphere transports heat energy absorbed by the Earth surface from the Sun to the Top Of Atmosphere (TOA) from where it is radiated to outer space, and thus acts as an air conditioner or heat engine[2] keeping the surface temperature constant under radiative forcing from the Sun. A basic question in climate science is the stability of this air conditioner under varying forcing, more specifically the change of surface temperature under doubled concentration of atmospheric CO_2 (from 0.028% to 0.056%) , referred to as *climate sensitivity*.

The heat is transported by the atmosphere in a combination of thermodynamics (turbulent convection and phase change in evaporation/condensation) and radiation—roughly 2/3 by thermodynamics and 1/3 by radiation. The thermodynamics involves positive radiative forcing balanced by evaporation at low latitudes/altitudes from a warm ocean causing warm air to rise-expand-cool including poleward motion followed by negative radiative forcing balanced by condensation at high latitudes/altitudes causing cool air to descend-contract-warm closing a thermodynamic cycle, as indicated in Figure 1, during polar winter.

---

[2] H. Osawa, A. Ohmura, R. D. Lorenz, T. Pujol, The Second Law of Thermodynamics and the Gobal Climate System: A Review of the Maximum Entropy Production Principle, Reviews of Geophysics, 41, 4 1018, 2003.

The Middle Atmosphere

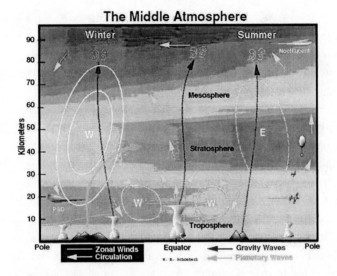

Figure 1: Thermodynamics of the atmosphere[3]

The Illusory Greenhouse Effect

The main message to the World and its leaders from the 2007 IPCC Fourth Assessment Report (AR4) is a prediction of an alarming climate sensitivity in the range 1.5-4.5°C, with a 'best estimate' of 3°C as a result of a so-called *greenhouse effect*.

The physics of this effect is claimed to have been identified and scientifically described by Fourier[4], Tyndall[5] and Arrhenius[6].

---

[3] NASA UARS Project

[4] Joseph Fourier, General remarks on the temperature of the earth and outer space. American Journal of Science. 32, 1-20 (1837) by Ebeneser Burgess. English translation of "Remarques ge´ne´rales sur les tempe´ratures du globe terrestre et des espaces plane´taires." Annales de Chimie et de Physique. (Paris) 2nd ser., 27, 136-67 (1824), by Jean-Baptiste Joseph Fourier.

[5] . John Tyndall, *On the Absorption and Radiation of Heat by Gases and Vapours, and on the Physical Connexion of Radiation, Absorption, Conduction.*-

Claes Johnson

An inspection of these sources shows very simplistic rudimentary analysis with only a simple model for radiation and no thermodynamics, which is the main message of this chapter; the mathematics of the Fourier-Tyndall-Arrhenius greenhouse effect is dead, and never was alive!

Further, to confuse the discussion, the 'greenhouse effect' is described with a misleading double-meaning...it is both the combined total effect of the atmosphere on the Earth's surface temperature including both radiation and thermodynamics, and at the same time, a hypothetical radiative effect of 'greenhouse gases' including CO_2 without thermodynamics.

In this way the 'greenhouse effect' becomes real, because it is the total effect of the atmosphere. The atmosphere—undeniably—has an effect, an 'atmosphere effect', while at the same time it can be linked to CO_2—apparently acting like a powerful 'greenhouse gas' capable of global warming upon a very small increase of 0.028%.

The simplest version of the 'greenhouse effect' is described by Stefan-Boltzmann's Law (SBL) $= \sigma T^4$, which in differentiated form appears as so:

$$dQ = \sigma 4T^3 dT = 4\frac{Q}{T} dT \sim 4dT$$

With $Q \approx 280W/m^2$ and $T \approx 288K$, this gives a climate sensitivity of about 1"C by attributing a certain fictitious additional 'radiative forcing' $dQ = 4W/m^2$ to doubled CO_2.

The Bakerian Lecture, The London, Edinburgh, and Dublin Philosophical Magazine and Journal of Science, Series 4, Vol. 22, pp. 169194, 273-285, 1861.

[6] Svante Arrhenius, *On the Influence of Carbonic Acid in the Air upon the Temperature of the Ground*, Philosophical Magazine and Journal of Science Series 5, Volume 41, April 1896, pages 237-276.

Since the total radiative forcing from the Sun is not assumed to change, the additional radiative forcing is supposed to result from a shift of the 'characteristic emission level/altitude' to a higher level at lower temperature caused by less radiation escaping to space from lower levels by increasing absorption by CO_2.

In this argument, the outgoing radiation is connected to a *lapse rate* (decrease of temperature with increasing altitude) supposedly being determined by thermodynamics. With lower 'characteristic emission temperature' at higher altitude, the whole temperature profile will have to be shifted upwards thus causing warming on the ground.

This is the starting point of the climate alarmism propagated by IPCC, a basic climate sensitivity of 1°C, which then is boosted to 3°C by various so-called positive 'feedbacks'.

The basic argument is: since Stefan-Boltzmann's Law cannot be disputed, and because CO_2 has certain properties of absorption/emission of radiation (light), which can be tested in a laboratory, the starting value of 1°C is an 'undeniable physical fact which cannot be disputed'. Even skeptics like Richard Lindzen and Roy Spencer accept it, and if skeptics believe something, then it must be true, right?

But wait!

Science does not work that way—science obeys facts and logical mathematical arguments, which are the essence of the scientific method. Let us now check if the basic postulate of a 'greenhouse effect' with basic climate sensitivity of 1°C can qualify as science.

Remember: climate politics without legitimate climate science is dead politics.

Mathematical Climate Simulation

The language and methodology of science, in particular climate science, is mathematics. Physical laws are expressed as differential equations of the principal form $D(u) = F$, where F represents *forcing*, u represents the corresponding *system state* coupled to F through a differential operator $D(u)$ acting on u.

With given forcing F, the corresponding state u can be determined by solving the differential equation $D(u) = F$. This is the essence of the scientific method. Note that the differential equation $D(u) = F$ usually describes a cause-effect relation in the sense that the system state u responds to a known given forcing F in a (stable) *forward problem*. This corresponds to putting the horse in front of the wagon, and not the other way around which is referred to as an (unstable) *inverse problem* with the state u given and F the forcing being sought.

Consider now the following approaches to modeling and simulating global climate:

A. Thermodynamics with radiative forcing (NSE with SBL forcing).

B. Radiation $dQ \sim 4dT$ as differentiated form of (SBL).

C. Radiation $dQ \sim 4dT$ combined with thermodynamic lapse rate.

D. Radiation $dQ \sim 4dT$ combined with thermodynamic lapse rate and feedback.

Here (A) is the (stable) forward problem described in the first section and studied below. (B) is self-referential without thermodynamics. (C-D) represent the IPCC approach as an (unstable) inverse problem of radiation with thermodynamic forcing with potentially large positive feedbacks and high climate sensitivity.

Altogether, (A) represents a stable forward problem compatible with a rational, scientific analysis, whereas the (C-D) of the IPPC represents an unstable inverse problem of questionable value.

In its popular form, the basic IPCC climate sensitivity of $1°C$ is claimed to come from a 'greenhouse gas'—the ability of CO_2 to 'trap heat', which is supposed to convince the uneducated. In its more elaborate form intended for the educated, it is connected to a thermodynamic lapse rate and characteristic emission level in order to account for an effect of additional radiative forcing without change of total radiative forcing. Both forms are severely simplistic and cannot be described as science.

To follow (A) we must rid ourselves of the common misconception of thermodynamics expressed in the quote above by Sommerfeld—that thermodynamics is beyond comprehension for mortals, in particular its Second Law. This is why climate scientists focus on radiation only, as something understandable, backing away from thermodynamics as something nobody can grasp. But it is possible to give thermodynamics and the Second Law a fully understandable meaning as I show in[7][8] and recall below. This insight opens a rational approach to climate dynamics, as (A) thermodynamics with radiative forcing.

Lapse Rate and Global Warming/Cooling

A theory is the more impressive the greater the simplicity of its premises, the more different kinds of things it relates to, and the more extended its area of applicability. This was therefore the deep impression that classical thermodynamics made upon me. It is the only physical theory of universal content which I

---

[7] J. Hoffman and C. Johnson, *Computational Turbulent Incompressible Flow*, Springer, 2007.
[8] Hoffman and C. Johnson, *Computational Thermodynamics*, http://www.nada.kth.se/ cgjoh/ambsthermo.pdf.

Claes Johnson

am convinced will never be overthrown, within the framework
of applicability of its basic concepts.
—Albert Einstein

The effective blackbody temperature of the Earth with atmosphere is -18°C, which can be allocated to a TOA at an altitude of 5km at a lapse rate of 6.5°C/km connecting TOA to an Earth surface at 15°C with a total warming of $5 \times 6.5 = 33$°C. The lapse rate determines the surface temperature since the TOA temperature is determined to balance a basically constant insolation.

What is then the main factor determining the lapse rate?

Is it explained by radiation or thermodynamics, or both?

Climate alarmism, as advocated by IPCC, is based on the assumption that radiation alone sets an initial lapse rate of 10°C/km, which then in reality is moderated by thermodynamics to an observed 6.5°C/km. Doubled CO_2 would then increase the initial lapse rate and with further positive thermodynamic feedback, the IPCC predicts an alarming climate sensitivity of 3°C—leading to dangerous global warming.

Climate alarmism skeptics like Lindzen and Spencer buy the argument of an initial rate of 10°C/km determined by radiation, but suggest that negative thermodynamic feedback effectively reduces climate sensitivity to a harmless 0.5°C.

We will argue that an initial lapse rate of $g = 9.81$ °C/km is instead determined by thermodynamics (and not by radiation) as an equilibrium state without heat transfer, which then in reality by thermodynamic heat transfer (turbulent convection/phase change) is decreased to the observed 6.5°C/km, with the heat transfer balancing the radiative heat forcing. More CO_2 would then require more heat transfer by thermodynamics and thus to a further decrease of the lapse rate rather than an increase. The atmosphere would then act like a boiling pot of water which under increased heating would boil more vigorously but not get any warmer.

Figure 2: The atmosphere maintains a constant surface temperature under increasing radiative heat forcing by increasing vaporization and turbulent convection, like a boiling pot of water on a stove.

In short: if thermodynamics is the main mechanism of the atmosphere as an air conditioner or heat transporter, then CO_2 will not cause warming, and the IPCC climate alarmism collapses.

We thus identify a basic difference between atmospheric heat transport by radiation (similar to conduction) and by thermodynamics of convection/phase change.

In radiation/conduction, increased heat transport couples to increased lapse rate (warming). In convection/phase change increased heat transport couples to decreased lapse rate (cooling).

Claes Johnson

Euler Equations for the Atmosphere

Every mathematician knows it is impossible to understand an elementary course in thermodynamics.
—Mathematician Vladimir Arnold

The viscosity of both water and air is small, while the spatial dimensions of the ocean and atmosphere are large, which means that the Reynolds number $R_e = \dfrac{UL}{v}$ is very large ($> 10^8$), where $U > 1\text{m/s}$ is a typical velocity, $L > 10^3\,\text{m}$ a length scale and $v < 10^{-5}$ a viscosity. Global climate thus results from turbulent flow at very large Reynolds numbers effectively in the form of turbulent solutions of the *Euler equations*[9].

We focus now on the atmosphere and as a model we consider the Euler equations for a compressible perfect gas occupying a volume Ω representing the troposphere, here for simplicity without Coriolis force from rotation:

Find (ρ, u, T) with ρ = density, u = velocity and T = temperature depending on x $\in \Omega$ and t> 0, such that for x $\in \Omega$ and t> 0:

$$D_u \rho + \rho \nabla \cdot u = 0,$$
$$D_u m + m \nabla \cdot u + \nabla p + g\rho e_3 = 0, \quad (1)$$
$$D_u T + RT \nabla \cdot u = q,$$

where: $m = \rho u$ (momentum), $p = R\rho T$ (pressure)

---

[9] J. Hoffman and C. Johnson, *Computational Turbulent Incompressible Flow*, Springer, 2007.

$R = c_p - c_v$ (c_v and c_p are specific heats under constant volume and pressure), and $D_u v = \dot{v} + u \cdot \nabla v$ is the material time derivative with respect to the velocity u with $\dot{v} = \frac{\partial v}{\partial t}$ the partial derivative with respect to time $t, e_3 = (0,0,1)$ is the upward direction, g gravitational acceleration and q is a heat source.

For air, $c_p = 1$ and $\frac{c_p}{c_v} = 1.4$. The Euler equations are complemented by initial values for ρ, m and T at $t = 0$, and the boundary condition $u \times n = 0$ on the boundary of Ω where n is normal to the boundary.

We assume that the heat source q adds heat energy at lower latitudes/altitudes and subtracts heat at higher latitudes/altitudes (radiation to outer space) including evaporation (subtraction of heat) at low altitudes and condensation (addition of heat) at higher altitudes.

We thus consider the full three-dimensional Euler/Navier-Stokes equations without any simplification of the vertical flow as in two-dimensional geostrophic flow or in hydrostatic approximation of vertical momentum balance, as a required feature of the next generation of climate models[10] not present in the current generation[11]. This is important because the heat transport involves both horizontal and vertical flow, i.e., ascending air at low latitudes and descending air at high latitudes, combined with high altitude poleward flow and low altitude flow towards the Equator.

---

[10] J. Slingo et al, Developing the next generation climate systems models: challenges and achievements, Phil. Trans. R. Soc. A 2009 367, 815-831, doi: 10.1098/rsta.2008.0207.
[11] D. Frierson, Climate Models,
http : //courses.washington.edu/pcc587/notes/587 9.pdf

Claes Johnson

The First and Second Laws of Thermodynamics

...no one knows what entropy is, so if you in a debate use this concept, you will always have an advantage.
—John von Neumann to Claude Shannon

We recall the Second Law of Thermodynamics[12]:

$$\dot{K} + \dot{P} = W - D, \dot{E} = -W + D + Q, \text{(2)}$$

where:

$$K(t) = \frac{1}{2}\int_\Omega \rho u \cdot u(x,t)dx, \ P(t) = \int_0^t \int_\Omega g\rho u(x,s)dxds,$$

$$E(t) = \int_\Omega c_v\rho T(x,t)dx, W(t) = \int_\Omega p\nabla \cdot u(x,t)dx,$$

$$Q(t) = \int_\Omega q(x,t)dx,$$
(3)

defines the momentary total kinetic energy $K(t)$, potential energy $P(t)$, internal energy $E(t)$ and work rate $W(t)$, and $D(t) \geq 0$ is rate of turbulent dissipation and $Q(t)$ rate of supplied heat or heat forcing. The work W is positive in expansion with $\nabla \cdot u$ positive, and negative in compression with $\nabla \cdot u$ negative (since the pressure p is positive).

Adding the two equations of the Second Law, we find that the change of total energy $(K + P + E)$ is balanced by the heat forcing:

$$\frac{d}{dt}(K + P + E) = Q, \text{(4)}$$

---

[12] Hoffman and C. Johnson, *Computational Thermodynamics*, http://www.nada.kth.se/ cgjoh/ambsthermo.pdf

which can be viewed to express the First Law of Thermodynamics as conservation of total energy.

Thermodynamics essentially concerns transformations between heat energy E and the sum $K + P$ of kinetic and potential energies with the transfer being $\pm(W - D)$: whatever $K + P$ gains is lost by E and vice versa. The Second Law sets the following limits for these transformations:

- heat energy E can be transformed to kinetic/potential energy $K + P$ only under expansion with $W > 0$;

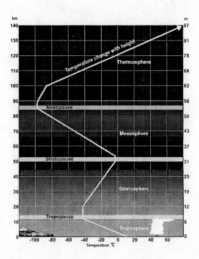

Figure 3: Temperature profile of the atmosphere, with constant lapse rate in the troposphere of 6.5°C/Km[13]

- turbulent dissipation D can transform kinetic/potential energy $K + P$ into heat energy E;
- turbulent dissipation D cannot transform heat energy to kinetic/potential energy, because $D \geq 0$.

---

[13] NOAA

Claes Johnson

Basic Isothermal and Isentropic Solutions

As anyone who has taken a course in thermodynamics is well aware, the mathematics used in proving Clausius' theorem (the Second Law) is of a very special kind, having only the most tenuous relation to that known to mathematicians.
—Mathematician Stephen Brush

We identify the following hydrostatic equilibrium base solutions, here fitted to an observed Earth surface temperature of 288 K, assuming $Q = 0$:

$$\bar{u} = 0, \bar{T} = 288 - gx_3,$$
$$\bar{\rho} = \alpha(288 - gx_3)^{\frac{1}{\gamma}},$$
$$\bar{p} = R\alpha(288 - gx_3)^{\frac{1}{\gamma}+1},$$
$$\bar{u} = 0, \bar{T} = 288(K),$$
$$\bar{\rho} = \alpha \exp(-gx_3),$$
$$\bar{p} = R288\alpha \exp(-gx_3),$$
$$(5)$$

where:

$$\gamma = \frac{R}{c_v} = 0.4$$

and thus:

$$R\left(\frac{1}{\gamma} + 1\right) = c_p = 1$$

where we scale x_3 in km and α denotes a positive constant to be determined by data.

The first solution is non-turbulent (or isentropic) with $D = 0$ in the Second Law:

$$\dot{E} + W = 0, \quad (6)$$

272

or in conventional notation

$$c_v dT + p dV = 0, \quad (7)$$

which combined with hydrostatic balance

$$\frac{\partial p}{\partial x_3} = -g\rho$$

and the differentiated form $p dV + V dp = R dT$ of the gas law, gives

$$(c_v + R)\frac{\partial T}{\partial x_3} = -g, \quad (8)$$

With $c_v + R = c_p = 1000 J/Kkg$, the heat capacity of dry air, we obtain an isentropic *dry adiabatic lapse rate* of $10°C/km$. With the double heat capacity of saturated moist air we obtain an isentropic *moist adiabatic lapse* of $5°C/km$.

The second solution has constant temperature and exponential drop of density and pressure, and can be associated with lots of turbulent dissipation (with $D = W$)—effectively equalizing the temperature. We summarize the properties of the above solutions (with $Q = 0$):

- isothermal: maximal turbulent dissipation: $D = W$,
- isentropic: minimal turbulent dissipation: $D = 0$.

We find real solutions between these extreme cases, with roughly $D = \frac{W}{2}$ and $\bar{\rho} \sim (288 - gx_3)^5$, $\bar{p} \sim (288 - gx_3)^6$, with a quicker drop with height than for the isentropic solution with $\bar{\rho} \sim (288 - gx_3)^{2.5}$ and $\bar{p} \sim (288 - gx_3)^{3.5}$, or turned the other way, with a smaller lapse rate of $6.5°C/km$.

Claes Johnson

Basic Thermodynamics

...thermodynamics is a dismal swamp of obscurity...a prime example to show that physicists are not exempt from the madness of crowds...Clausius' verbal statement of the second law makes no sense...All that remains is a Mosaic prohibition; a century of philosophers and journalists have acclaimed this commandment; a century of mathematicians have shuddered and averted their eyes from the unclean...Seven times in the past thirty years have I tried to follow the argument Clausius offers and seven times has it blanked and gravelled me. I cannot explain what I cannot understand.
—Physicist Clifford Truesdell

We have formulated a basic model of the atmosphere acting as an air conditioner/refrigerator by transporting heat energy from the Earth surface to the top of the atmosphere in a thermodynamic cyclic process with radiation/gravitation forcing, consisting of:
  • ascending/expanding/cooling air heated by low altitude/latitude radiative forcing;
  • descending/compressing/warming air cooled by high altitude/latitude outgoing radiation,
combined with low altitude evaporation and high altitude condensation.

The model is compatible with observation and suggests that the lapse rate/surface temperature is mainly determined by thermodynamics and not by radiation.

The thermodynamics of a standard refrigerator requires a compressor, which in the case of an atmosphere is taken over by gravitation causing compression of descending air.

EARTH'S ENERGY BUDGET

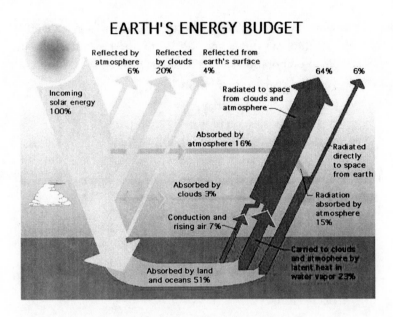

Figure 4: Earth's Energy Budget[14]

Basic Data

You can fool all the people some time, and some of the people all the time, but you cannot fool all the people all the time.
—Abraham Lincoln

We collect the following observed data, for the first half of the above cycle:

- average upward velocity = 0.01 m/s;
- average density = 0.6kg/m^3 ;
- average altitude of TOA = 5000m;
- c_p = 1000 J/Kkg

---

[14] NASA Atmospheric Science Data Center

- $Q \approx 180\text{W/m}^2$ absorbed by the Earth surface with 60W allocated to radiation, and 120W to thermodynamics with 100W to evaporation and 20W to convection.
- observed lapse rate $\approx -6.5°\text{C/km}$,
- evaporation $\approx 4\text{cm/day}$,
- heat of vaporization of water 2200 kJ/kg,
- turbulent dissipation rate: 0.002 W/kg,

For the upward motion of a column of air over a square meter of surface, we have:

- $\dot{P} \approx 0.01 \times 0.7 \times 5000 \times g = 350\text{W}$,
- $\dot{E} \approx -0.01 \times 0.7 \times 1000 \times 5000 \times \dfrac{6.5}{1000} \approx -230 \text{ W}$,
- phase change: $2.2 \times 10^6 \times 10^2 \times \dfrac{0.04}{24 \times 3600} \approx 100\text{W}$,

which is compatible with:

$$W - D = \dot{P} = 350W$$

and

$$\dot{E} = -W + D + Q = -230\,W.$$

The observed lapse rate of 6.5°C/km can be viewed as being obtained by moderating the dry adiabatic rate of 9.8°C/km by a combined process of phase change and turbulent dissipation effectively reducing the drop of temperature with altitude. The energy transfer in this process is approximately:

$$\frac{3.5}{6.5} \times 230 = 120W,$$

with 100-110W for evaporation and

$$20 = 0.002 \times 5000,$$

approximately 10-20W for turbulence, which is roughly equal to the heat forcing allocated to thermodynamics (120W). Increasing heat transfer then corresponds to non-increasing lapse rate and nonwarming; the main message of our analysis.

The observed lapse rate of 6.5°C/km is bigger than the moist adiabatic rate of 5°C/km, which causes unstable overturning of rising warm air and turbulent dissipation.

Lapse Rate vs. Radiative Forcing

If the lapse rate is L then $\dot{P} + \dot{E} = Q$, combined with $\dfrac{\dot{E}}{\dot{P}} = \dfrac{L}{10}$ according to the above computation, gives:

$$L = 10\left(1 - \frac{Q}{\dot{P}}\right).$$

If Q is increased, then L will decrease if $\dot{P}$ stays constant, but if $\dot{P}$ increases quicker than Q, then L may increase. Increasing Q may be expected to give an increase of $\dot{P}$ by increasing the vertical convection velocity, but a decrease by increasing phase change evaporation/condensation. Which effect will dominate? Convection or phase change? Computations with an answer are under way...until then we notice that out of 120 W/m^2 of radiative heat forcing, a major part—of say 100—can be allocated to phase change, which gives phase change a good chance to compete with convection...

Summary: Atmosphere as Air Conditioner

A good many times I have been present at gatherings of people who, by the standards of the traditional culture, are thought highly educated and who have with considerable gusto been expressing their incredulity at the illiteracy of scientists. Once or twice I have been provoked and have asked the company how many of them could describe the Second Law of Thermodynamics. The response was cold: it was also negative.
—C. P. Snow in 1959 Rede Lecture entitled *The Two Cultures and the Scientific Revolution*

Let us now sum up the experience from our analysis. We have seen that the atmosphere acts as a thermodynamic air conditioner transporting heat energy from the Earth's surface to a TOA under radiative heat forcing. We start from an isentropic stable equilibrium state with lapse rate 9.8°C/km with zero heat forcing and discover the following scenario for the response of the air conditioner under increasing heat forcing:

1. increased heat forcing of the ocean surface at low latitudes is balanced by increased vaporization;
2. increased vaporization increases the heat capacity which decreases the moist adiabatic lapse rate;
3. if the actual lapse rate is bigger than the actual moist adiabatic rate, then unstable convective overturning is triggered;
4. unstable overturning causes turbulent convection with increased heat transfer.

The atmospheric air conditioner thus may respond to increased heat forcing by increased vaporization decreasing the moist adiabatic lapse rate combined with increased turbulent convection if the actual lapse rate is bigger than the moist adiabatic lapse rate. This is how a boiling pot of water reacts to increased heating.

Climate Thermodynamics

If someone points out to you that your pet theory of the universe is in disagreement with Maxwell's equations, then so much the worse for Maxwell's equations. If it is found to be contradicted by observation, well, these experimentalists do bungle things sometimes. But if your theory is found to be against the second law of thermodynamics, I can give you no hope; there is nothing for it but to collapse in deepest humiliation.

—Sir Arthur Stanley Eddington in *The Nature of the Physical World*, 1915

Chapter 19

Computational Blackbody Radiation
by Claes Johnson

Blackbody Radiation

All these fifty years of conscious brooding have brought me no nearer to the answer to the question, "What are light quanta?". Nowadays every Tom, Dick and Harry thinks he knows it, but he is mistaken.
—Albert Einstein (1954)

Wave-Particle Duality and Modern Physics

MAXWELL'S EQUATIONS REPRESENT a culmination of classical mathematical physics by offering a compact mathematical formulation of all of electromagnetics including the propagation of light and radiation, as electromagnetic waves. But, like in a Greek tragedy, the success of Maxwell's equations prepared classical mathematical physics for collapse and the rise of modern physics based on a concept *wave-particle duality*—with a resurrection of Newton's old idea of light as a stream of light particles or photons, combined with statistics in its modern version.

281

Claes Johnson

But elevating wave-particle duality into a physical principle is a cover-up of a contradiction.[1] [2] [3] As a reasonable human being, you might sometimes act like a fool, but here, duality can be called schizophrenia, and schizophrenic science is crazy science—in our time represented by CO_2 climate alarmism, which is ultimately based on viewing radiation as streams of particles.

The purpose of this chapter is to show that particle statistics can be replaced by deterministic finite-precision computational wave mechanics. We thus seek to open a door to restoring reason to physics—including climate physics—without contradictory wave-particle duality.

Climate Alarmism, Greenhouse Effect and Backradiation

In particular, the objective is to show that the 'greenhouse effect' of climate alarmism claimed to arise from 'backradiation' of

---

[1] *I consider it quite possible that physics cannot be based on the field concept, i.e., on continuous structures. In that case, nothing remains of my entire castle in the air, gravitation theory included, and of the rest of physics.*
—Albert Einstein (1954)

[2] *What I wanted to say was just this: in the present circumstances the only profession I would choose would be one where earning a living had nothing to do with the search for knowledge.*
—Einstein's last letter to Max Born Jan 17, 1955 shortly before his death on the 18th of April, probably referring to Born's statistical interpretation of quantum mechanics.

[3] Schrodinger, *The Interpretation of Quantum Physics*, Ox Bow Press, Woodbridge, CN, 1995: *What we observe as material bodies and forces are nothing but shapes and variations in the structure of space. Particles are just schaumkommen (appearances). ...let me say at the outset, that in this discourse, I am opposing not a few special statements of quantum physics held today (1950s), I am opposing as it were the whole of it, I am opposing its basic views that have been shaped 25 years ago, when Max Born put forward his probability interpretation, which was accepted by almost everybody.*

particle streams as depicted by NASA in Figure 4, is pure fiction without real physical meaning. This removes a main source of energy from climate alarmism, in the sense that various feedbacks will have to start from zero rather than creating alarming warming from radiation alone. We first give a popular science description in words and then a mathematical one using formulas.

To express physics in precise terms it is necessary to use the language of mathematics, but the main ideas can be captured in ordinary language—leading to understanding. The two forms of expression complement each other. In particular, we shall find that the term 'backradiation', which can be contemplated without mathematics, when expressed mathematically, reveals its true unstable nature, which reveals it as a reality-free, fictitious, unphysical phenomenon. We shall find that it represents the same form of fiction as a bubble-economy in real economic terms— fictitious values without real substance fueled by a circulating, self-propelling flow of paper money.

Blackbody Radiation in Words

A blackbody acts like a transformer of radiation; it absorbs high-frequency radiation and emits low-frequency radiation. The temperature of the blackbody determines a *cut-off frequency* for the emission, which increases linearly with the temperature. The warmer the blackbody is, the higher frequencies it can and will emit. While all frequencies are being absorbed, only frequencies below cut-off are emitted.

A blackbody thus can be seen as a system of resonators with different Eigen-frequencies which are excited by incoming radiation and then emit radiation. An ideal blackbody absorbs all incoming radiation and remits all absorbed radiation below cut-off.

Claes Johnson

Conservation of energy requires absorbed frequencies above cut-off to be stored in some form, more precisely as heat energy—thus increasing the temperature of the blackbody.

As a transformer of radiation, a blackbody acts in a very simple way: it absorbs all radiation, emits absorbed frequencies below cut-off, and uses absorbed frequencies above cut-off to increase its temperature. A blackbody thus acts as a semi-conductor transmitting only frequencies below cut-off, and transforms coherent frequencies above cut-off into heat in the form of incoherent, high-frequency noise.

Here, we distinguish between coherent organized electro-magnetic waves of different frequencies in the form of radiation or light, and incoherent high-frequency vibrations or noise, perceived as heat.

A blackbody absorbs and emits frequencies below cut-off without getting warmer, while absorbed frequencies above cut-off are not emitted, but are instead stored as heat energy—increasing its temperature.

A blackbody is like an amplifier with a restricted range of frequencies, or low-pass filter, which remits/amplifies frequencies below a cut-off frequency and dampens frequencies above cut-off with the damped wave energy being turned into heat.

A blackbody acts like a censor which filters out coherent high-frequency (dangerous) information by transforming it into incoherent (harmless) noise. The IPCC acts like a blackbody by filtering coherent critical information—transforming it into incoherent nonsense perceived as hazardous global warming.

The increase of the cut-off frequency with temperature can be understood as an increasing ability to emit coherent waves with increasing temperature/excitation or wave amplitude. At low temperature, waves of small amplitude cannot carry a high-pitched signal. It is like speaking at -40°C with very stiff lips.

284

Computational Blackbody Radiation

We can also compare this mechanism with a common teacher-class situation where an excited, high temperature teacher emits information over a range of frequencies from low (simple material) to high (difficult material), which the class absorbs, re-emits and repeats—below a certain cut-off frequency, while the class is unable to emit/repeat frequencies above cut-off, which is used to increase the temperature or frustration/interest of the class. The temperature of the class can never exceed the temperature of the teacher, because all coherent information originates from the teacher. The teacher and student connect in two-way communication with a one-way flow of coherent information[4].

The net result is that a warm blackbody can heat a cold blackbody, but not the other way around. A teacher can teach a student but not the other way around[5]. The hot Sun heats the colder Earth, but the Earth does not heat the Sun. The Earth's warm surface can heat a cold atmospheric layer, but a cold atmosphere cannot heat the Earth's warm surface. A blackbody is heated only by frequencies which it cannot emit, stored as heat energy.

There is no 'backradiation' from the atmosphere to the Earth. There is no 'greenhouse effect' from 'backradiation. Figure 4, propagated by NASA, displays fictional non-physical recirculating radiation with an Earth surface emitting 117%, while absorbing 48% from the Sun.

We shall see that recirculation of energy is nonphysical— because it is unstable. The instability is of the same nature as that of an economy with income tax approaching 100%, or an interest rate approaching 0%, or benefits without limits from increasing taxes without limit. An economy with fictitious money circling with

---

[4] Publisher's note: here, the professor has a little fun at we student's expense. We forgive him, he's earned the right.
[5] Op cit.

increasing velocity creates financial bubbles which burst sooner or later from inherent instability, as we witnessed in recent times.

An atmosphere with recirculating radiation would be unstable and thus cannot exist over time.

There is no 'backradiation' for the same reason as there is no 'backconduction' or 'backdiffusion', namely instability. 'Backdiffusion' would correspond to restoring a blurred, diffused image using Photoshop, which you can easily convince yourself is impossible. Take a sharp picture and blur it, and then try to restore it by sharpening—you will discover this does not work, because of instability. Blurring or diffusion destroys fine details— which are impossible to recover. Diffusion or blurring is like taking mean values of individual data samples—individual data samples cannot be recovered from mean values. Mixing milk into your coffee by stirring/blurring is possible, but unmixing is impossible by unstirring/unblurring.

Radiative heat can be transmitted by electromagnetic waves from a warm blackbody to a colder blackbody, but not from a cold to a warmer, thus we have a one-way direction of heat energy, while electromagnetic waves propagate in both directions. We thus distinguish between the two-way propagation of waves and the one-way propagation of heat energy by waves.

A cold body can heat up by eating/absorbing high-frequency, high-temperature coherent waves in a catabolic process of destruction of coherent waves into incoherent heat energy. A warm body cannot heat up by eating/absorbing low-frequency low-temperature waves, because catabolism involves destruction of structure. Anabolism builds structure, but a blackbody is only capable of destructive catabolism (the metabolism of a living cell consists of destructive catabolism and constructive anabolism).

Planck's Law

The particle nature of light of frequency v as a stream of *photons* of energy hv with h Planck's constant, is supposed to be motivated by Einstein's model of the photoelectric effect[6], viewed to be impossible[7][8] to explain assuming light is an electromagnetic wave phenomenon satisfying Maxwell's equations. The idea of light in the form of energy quanta of size hv was introduced by Planck[9] in 'an act of despair' to explain the *radiation energy* $Rv(T)$ emitted by a *blackbody* as a function of frequency v and temperature T, per unit frequency, surface area, viewing solid angle and time:

$$R_v(T) = \gamma T v^2 \theta(v,T), \gamma = \frac{2k}{c^2}, \quad (1)$$

with the *high-frequency cut-off* factor:

$$\theta(v,T) = \frac{\frac{hv}{kT}}{e^{\frac{hv}{kT}} - 1}, \quad (2)$$

where c is the speed of light in vacuum, k is Boltzmann's constant, with $\theta(v,T) \approx 0$ for $\frac{hv}{kT} > 10$ say and $\theta(v,T) \approx 1$ for $\frac{hv}{kT} < 1$. Since $\frac{h}{k} \approx 10^{-10}$ this effectively means that only frequencies $v \leq T10^{11}$ will be emitted, which fits with the

---

[6] Einstein, A., *On a Heuristic Point of View Toward the Emission and Transformation of Light*, Ann. Phys. 17, 132, 1905.

[7] *If a scientist says that something is possible he is almost certainly right, but if he says that it is impossible he is probably wrong.*

—Arthur C. Clarke

[8] Thomas Kuhn, *Black-Body Theory and the Quantum Discontinuity*, 1894-1912, Oxford University Press 1978.

[9] Max Planck, *Acht Vorlesungen über Theoretische Physik, Füfte Vorlesung: Wärmestrahlung und Elektrodynamische Theorie*, Leipzig, 1910.

Claes Johnson

common experience that a black surface heated by the high-frequency light from the Sun, will not itself shine like the Sun, but radiate only lower frequencies. We refer to $\frac{kT}{h}$ as the *cut-off* frequency, in the sense that frequencies $v > \frac{kT}{h}$ will be radiated subject to strong damping. We see that the cut-off frequency scales with T, which is *Wien's Displacement Law*. In Figure 1, notice that the cut-off shifts to higher frequency with higher temperature according to Wien's Displacement Law.

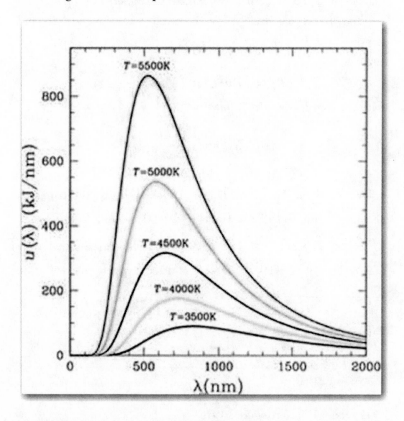

Figure 1: Radiation energy vs. wave length/frequency at different temperatures of a radiating blackbody.

Computational Blackbody Radiation

The term *blackbody* is conventionally used to describe an idealized object which absorbs all electromagnetic radiation falling on it, hence appearing to be black. The analysis to follow will reveal some of the real truth of a blackbody, such as the Earth, radiating infrared light while absorbing light from the Sun—mainly in the visible spectrum.

It is important to note that the constant $\gamma = \frac{2k}{c^2}$ is very small: with $k \approx 10^{-23} \frac{J}{K}$ and $c \approx 3x10^8 m/s$, we have $\gamma \approx 10^{-40}$. In particular, $\gamma v^2 << 1$ if $v \leq 10^{18}$ including the ultraviolet spectrum, a condition we will meet below.

By integrating and summing frequencies in Planck's radiation law (1), one obtains *Stefan-Boltzmann's Law* stating that the total radiated energy $R(T)$ per unit surface area emitted by a blackbody is proportional to T^4:

$$R(T) = \sigma T^4 \quad (3)$$

where $\sigma = \frac{2\pi^5 K^4}{15c^2 h^3} = 5.67 \times 10^{-8} W^{-1} m^{-2} K^{-4}$ is *Stefan-Boltzmann's constant.*

On the other hand, the classical *Rayleigh-Jeans Radiation Law* $R_v(T) \sim \gamma T v^2$ without the cut-off factor, results in an 'ultraviolet catastrophe' with infinite total radiated energy, since: $\gamma T \int_1^n v^2 dv \sim \gamma T n^3 \to \infty$ as $n \to \infty$. Here and below, we use $\sim$ to denote proportionality with the constant of proportionality close to 1.

Stefan-Boltzmann's Law fits reasonably well to observation, while the Rayleigh-Jeans Law leads to an absurdity—and so must, somehow, be incorrect. The Rayleigh-Jeans Law was derived from viewing light as electromagnetic waves governed by Maxwell's equations, which forced Planck in his 'act of despair' to give up the wave model and replace it with statistics of 'quanta'—viewing light as a stream of particles or photons. But the scientific cost of abandoning the wave model is very high, and we now present an alternative way of avoiding the catastrophe by modifying the wave model by *finite precision computation*, instead of resorting to particle statistics.

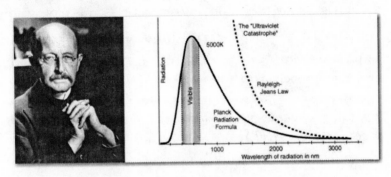

Figure 2: The Ultraviolet Catastrophe[10]

We shall see that finite precision computation introduces a high-frequency cut-off in the spirit of the finite precision computational model for thermodynamics[11].

---

[10] Planck on the ultraviolet catastrophe in 1900: ...the whole procedure was an *act of despair* because a theoretical interpretation had to be found at any price, no matter how high that might be...either the quantum of action was a fictional quantity, then the whole deduction of the radiation law was essentially an illusion representing only an empty play on formulas of no significance, or the derivation of the radiation law was based on sound physical conception.
Planck in 1909: Mechanically, the task seems impossible, and we will just have to get used to it (quanta).
[11] J. Hoffman and C. Johnson, *Computational Thermodynamics*, http://www.nada.kth.se/cgjoh/ambsthermo.pdf

The scientific price of resorting to statistical mechanics is high, as was clearly recognized by Planck and Einstein, because the basic assumption of statistical mechanics of microscopic games of roulette seems both scientifically illogical and impossible to verify experimentally. Thus, statistical mechanics runs the risk of representing *pseudo-science* because of obvious difficulties of testing the fundamental assumptions.

The purpose of this note is to present an alternative to particle statistics for blackbody radiation based on deterministic, finite precision computation in the form of *General Galerkin G2*[12] [13].

To observe individual photons as 'particles' without both mass and charge seems impossible, and so the physical reality of photons has remained hypothetical with regard to explaining blackbody radiation and the photoelectric effect. If explanations can be given by wave mechanics, then both the contradiction of wave-particle duality and the mist of statistical mechanics can be avoided, thus fulfilling a dream of the late Einstein[i] [ii].

The Enigma

The basic enigma of blackbody radiation can be given different formulations:

- Why is a blackbody black (invisible), emitting infrared radiation when 'illuminated' by light in the visible spectrum?
- Why is radiative heat transfer between two bodies always directed from the warmer body to the colder?
- How can high-frequency radiation transform into heat energy?

---

[12] J. Hoffman and C. Johnson, *Computation Turbulent Incompressible Flow*, Springer 2008.
[13] J. Hoffman and C. Johnson, *Computational Thermodynamics*, http://www.nada.kth.se/cgjoh/ambsthermo.pdf

- Why does heat energy transform to radiation of a certain frequency only if the temperature is high enough?

We shall find the answer in analyzing *resonance in a system of oscillators* (oscillating molecules/charges):

- incoming radiation is absorbed by resonance;
- absorbed incoming radiation is emitted as outgoing radiation, or is stored as internal heat energy;
- outgoing radiation has a frequency spectrum $\sim \gamma T \nu^2$ for $\nu \lesssim T$, assuming all frequencies ν have the same temperature T, with a cut-off to zero for $\nu \gtrsim T$;
- incoming frequencies below cut-off are emitted;
- incoming frequencies above cut-off are stored as internal heat energy.

Waves vs. Particles in Climate Science

We shall find answers to these questions using a wave model where we can separate propagation of waves and propagation of heat energy by waves. This allows two-way propagation of waves and one-way propagation of heat energy. In a particle model this separation is impossible since the heat energy is tied to the particles.

Radiation as a stream of particles leads to an idea of 'backradiation' with two-way propagation of heat energy carried by a two-way propagation of particles. We argue that such two-way propagation is unstable because it requires cancellation, and cancellation in massive two-way flows of heat energy is unstable to small perturbations and thus is unphysical.

We find the supposed scientific basis of climate alarmism is unstable and therefore will collapse under perturbations, even small ones, with Climategate representing a perturbation which is big rather than small...

Wave Equation with Radiation

There are no quantum jumps, nor are there any particles.
—Physicist Heinz-Dieter Zeh[14]

Basic Radiation Model

For simplicity, we consider the wave equation with radiation in a one-space dimension with assumed periodicity:

Find $u = u(x, t)$ such that:

$$\ddot{u} - u'' - \gamma \dddot{u} = f, \; -\infty < x, t < \infty \quad (4)$$

where (x, t) are space-time coordinates, $\dot{u} = \dfrac{du}{dt}$, $u' = \dfrac{du}{dx}$, $f(x, t)$ models forcing in the form of incoming waves, and the term $-\gamma \dddot{u}$ models outgoing radiation with $\gamma > 0$ a small constant.

This model, in the spirit of Planck[15], before collapsing to the statistics of quanta, describes a continuous string of vibrating charges absorbing energy with forcing f of intensity f^2 and radiating energy of intensity $\gamma \ddot{u}^2$. The radiation term has the form $-\gamma \dddot{u} \sim \dot{F}$, where $F \sim \ddot{u}$ represents the electrical field generated by an oscillating charge at position x with acceleration $\ddot{u}(x, t)$.

Basic Energy Balance

Multiplying (4) by $\dot{u}$ and integrating by parts over a space period, we obtain:

---

[14] H.D. Zeh, Physics Letters A 172, 189-192, 1993.

[15] Max Planck, *Acht Vorlesungen über Theoretische Physik, Füfte Vorlesung: Wärmestrahlung und Elektrodynamische Theorie*, Leipzig, 1910.

Claes Johnson

$$\int (\ddot{u}\dot{u} + \dot{u}'u')dx + \int \gamma\ddot{u}^2 dx = \int f\dot{u}dx,$$

which we can write:

$$\dot{E} = A - R \quad (5)$$

where:

$$E(t) \equiv \frac{1}{2}\int -(\dot{u}(x,t)^2 + u'(x,t)^2 dx \quad (6)$$

is the internal energy viewed as heat energy, and

$$A(t) = \int f(x,t)\dot{u}(x,t)dx \,, R(t) = \int \gamma\ddot{u}(x,t)^2 dx \quad (7)$$

is the absorbed and radiated energy, respectively, with the difference $A - R$ driving changes of internal energy E.

If the incoming wave is an emitted wave: $f = -\gamma\ddot{U}$ with amplitude U, then:

$$\dot{E} = \int (f\dot{u} - \gamma\ddot{u}^2)dx = \int \gamma(\ddot{U}\ddot{u} - \ddot{u}^2)dx \le \frac{1}{2}(R_{in} - R)$$
$$(8)$$

with $R_{in=}\int \gamma\ddot{U}^2\, dx$ the incoming radiation energy, and R the outgoing.

We conclude that if $\dot{E} \ge 0$ then $R \le R_{in}$, that is, in order for energy to be stored as internal heat energy, it is required that the incoming radiation energy be bigger than the outgoing.

Of course, this is what is expected from conservation of energy. It can also be viewed as a Second Law of Radiation stating that radiative heat transfer is possible only from warmer to cooler. We shall see this basic law expressed differently more precisely below.

The Rayleigh-Jeans Radiation Law

But the conception of localized light-quanta out of which Einstein got his equation must still be regarded as far from established. Whether the mechanism of interaction between ether waves and electrons has its seat in the unknown conditions and laws existing within the atom, or is to be looked for primarily in the essentially corpuscular Thomson-Planck-Einstein conception of radiant energy, is the all-absorbing uncertainty upon the frontiers of modern Physics.
—Robert A. Millikan[16]

Spectral Analysis of Radiation

We shall show that the Rayleigh-Jeans radiation law:

$$R_v(T) = \gamma T v^2$$

is a direct consequence of the form of the radiation term $-\gamma \ddot{u}$, assuming that all frequencies have the same temperature T. This is elementary.

We shall also show that if the intensity of the forcing f in the model (4) has a Rayleigh-Jeans spectrum $\sim \gamma T v^2$, then so has the corresponding radiation energy $R_v(T)$. More precisely, we show as a main result that (the bar denotes integration in time):

$$R_v \sim \overline{f_v^2} \quad (9)$$

This is less elementary and results from a quite subtle phenomenon of near resonance.

---

[16] Robert A Millikan, *The electron and the light-quanta from the experimental point of view*, Nobel Lecture, May 23, 1923.

Claes Johnson

To prove this, we first make a spectral decomposition in x, assuming periodicity with period 2π:

$$\ddot{u}_v + v^2 u_v - \gamma \ddot{u}_v = f_v, \quad -\infty < t < \infty, v = 0, \pm1, \pm2, \ldots \tag{10}$$

into a set of damped linear oscillators with:

$$u(x,t) = \sum_{v=-\infty}^{\infty} u_v(t) e^{ivx}$$

We then use the Fourier transformation in t,

$$u_v(t) = \int_{-\infty}^{\infty} u_{v,\omega} e^{i\omega t} \, d\omega, \qquad u_{v,w} = \frac{1}{2\pi} \int_{-\infty}^{\infty} u_v(t) e^{-i\omega t} \, dt$$

to get, assuming u_v^3 can be replaced by $-v^2 \dot{u}_v$

$$(-\omega^2 + v^2) u_{v,\omega} + i\omega\gamma v^2 u_{v,\omega} = f_{v,\omega}.$$

We have by Parseval's formula:

$$\overline{u_v^2} \equiv \int_{-\infty}^{\infty} |u_v(t)|^2 dt = 2\pi \int_{-\infty}^{\infty} |u_{v,w}|^2 d\omega$$

$$= 2\pi \int_{-\infty}^{\infty} \frac{|f_{v,\omega}|^2 d\omega}{(v-\omega)^2(v+w)^2 + \gamma^2 v^4 \omega^2}$$

$$\sim \frac{1}{v^2} \int_{-\infty}^{\infty} \frac{|f_{v,\omega}|^2 d\omega}{(v-\omega)^2 + \gamma^2 v^4}$$

$$\sim \frac{1}{\gamma v^4} \int_{-\infty}^{\infty} \frac{|f_{v,v} + \gamma v^2 \overline{\omega}|^2 d\overline{\omega}}{\overline{\omega}^2 + 1}$$

296

where we used the change of integration variable $\omega = v + \gamma v^2 \bar{\omega}$.

We now assume that $|f_{v,v+\gamma v^2\bar{\omega}}|^2 \sim \overline{f_v^2}$ for $|\bar{\omega}| \leq 1$, which means that frequencies ω with $|v - \omega| \lesssim \gamma v^2$ contribute more or less equally to the excitation of the frequency v, because the resonance term $(v - \omega)^2$ then is dominated by the radiation term $\gamma^2 v^4$. This means that the radiation term acts like diffusion effectively blurring the ω-reading of the forcing $f_{v,\omega}$. With this assumption we get:

$$\overline{u_v^2} \sim \frac{1}{\gamma v^4} \overline{f_v^2}$$

that is,

$$R_v \equiv \overline{\gamma \ddot{u}_v^2} \approx \gamma v^4 \overline{u_v^2} = \gamma T_v v^2 \sim \overline{f_v^2}, \quad (11)$$

where $R_v = R_v(T_v)$ is the intensity of the radiated wave of frequency v, and we view

$$T_v = \frac{1}{2}(\overline{\dot{u}_v^2} + v^2 \overline{u_v^2}) \approx \overline{\dot{u}_v^2}$$

as the temperature of the corresponding frequency.

We read from (11) that

$$R_v(T_v) \approx \gamma T_v v^2 \quad (12)$$

which is the Rayleigh-Jeans Law. Further, if $\overline{f_v^2} \sim \gamma T v^2$, then also $R_v(T_v) \sim \gamma T_v^2$ with $T_v \sim T$. The emitted radiation will thus mimic an incoming Rayleigh-Jeans spectrum in temperature equilibrium with $T_v \sim T$ for all frequencies v.

Claes Johnson

We note the constant of proportionality in $R_v \sim \overline{f_v^2}$ independent of γ and V which reflects that the string has a certain absorbitivity (greater or equal to its emissivity).

Summing over frequencies we get:

$$R \equiv \frac{1}{2\pi} \int_0^{2\pi} \gamma \overline{\dot{u}^2} \, dx \sim \frac{1}{2\pi} \int_0^{2\pi} \overline{f^2} dx = \|f\|^2, \ (13)$$

that is, the intensity of the total outgoing radiation R is proportional to the intensity of the incoming radiation as measured by $\|f\|^2$, and thus $R \sim \|f\|^2$. We summarize this in Theorem 1.

Theorem 1

The radiation $R_v = \overline{\gamma \dot{u}_v^2}$ *of the damped oscillator (10) with forcing* f_v *satisfies* $R_v \sim \overline{f_v^2}$, *or after summation* $R \sim \|f\|^2$. *In particular, if* $\overline{f_v^2} \sim \gamma T v^2$ *then* $R_v = R_v(T_v) \sim \gamma T v^2$ *with* $T_v = T$.

Radiation from Near-Resonance

We have seen radiation resulting from forcing by a phenomenon of near-resonance in a damped oscillator of the form:

$$\ddot{u}_v + v^2 u_v + \gamma v^2 \dot{u}_v = f_v \ (14)$$

where the forcing f_v is balanced by the dynamics of the oscillator $\ddot{u}_v + v^2 u_v$ and the radiator $\gamma v^2 \dot{u}_v$ with an effect of dissipative damping (with $\gamma v^2 \leq 1$). In the case of large damping with $\gamma v^2 \approx 1$, then f_v is mainly balanced by the radiator, that is, $\gamma v^2 \dot{u}_v \approx \dot{u}_v \approx f_v$ with the result that $R_v = \overline{f_v \dot{u}_v} \approx \overline{f_v^2}$. We see that in this case $\dot{u}_v$ is *in-phase* with the forcing f_v, and there is little resonance with the oscillator.

We next consider the case $\gamma v^2 << 1$ with small damping and thus near-resonance. The relation $R_v = \overline{f_v \dot{u}_v} \sim \overline{f_v^2}$ tells us

298

that this case f_v is balanced by the dynamics of both the oscillator and the radiator with u_v in-phase and thus $\dot{u}_v$ *out-of-phase*. This is because, if not, then $\gamma v^2 \dot{u}_v \approx f_v$ with $\dot{u}_v$ in-phase, would lead to the contradicting $R_v = \overline{f_v \dot{u}_v} \sim \dfrac{\overline{f_v^2}}{\gamma v^2} \gg \overline{f_v^2}$.

Absorption vs. Emission

In the wave model (4) we associated the term $-\gamma \ddot{u}$ with radiation, but if we just read the equation, we only see a dissipative term absorbing energy without information about how this energy is dispensed with, e.g. by being radiated away. The model thus describes *absorption by* the vibrating string under forcing, and as written—not the process of *emission from* the string.

However, if we switch the roles of f and $-\gamma \ddot{u}$ and view $-\gamma \ddot{u}$ as input, then we can view f as an emitted wave, which can act as forcing on another system. For frequencies with $\gamma v^2 \ll 1$, we will then have:

$$\overline{f_v^2} \sim \gamma \overline{\ddot{u}_v^2} \gg \overline{(\gamma \ddot{u})^2} \approx \gamma v^2 \gamma \overline{\ddot{u}_v^2}$$

with the resulting emission boosted by resonance, as in the resonant amplification of a musical instrument (e.g the body of a guitar).

In both cases, the relation $R_v \sim \overline{f_v^2}$ expresses this: that the energy of the incoming absorbed radiation is equal to the outgoing (emitted) radiation.

Planck's Radiation Law

Would it not be possible to replace the hypothesis of light

Claes Johnson

quanta by another assumption that would also fit the known phenomena? If it is necessary to modify the elements of the theory, would it not be possible to retain at least the equations for the propagation of radiation and conceive only the elementary processes of emission and absorption differently than they have been until now?
—Albert Einstein

The Alexander Cut-Off by Planck

The Rayleigh-Jeans Law leads to an 'ultraviolet catastrophe' because, without some form of high-frequency limitation, the total radiation will be unbounded. Classical wave mechanics thus appears to lead to an absurdity, which must be resolved one way or the other.

In an 'act of despair' Planck escaped the catastrophe by cutting the Gordian Knot—simply replacing classical wave mechanics with a new statistical mechanics where high frequencies were assumed to be rare; 'a theoretical interpretation had to be found at any price, no matter how high that might be...'. It is like kicking out a good old horse which served many purposes, just because it has a tendency to 'go to infinity' with a certain stimulus, and then replacing it with a completely new wild horse which you don't understand and cannot control.

The price of throwing out classical wave mechanics is very high, and thus it is natural to ask if this is really necessary. Can we conceive a form of classical mechanics without the ultraviolet catastrophe? Can a cut-off of high frequencies be performed without a Gordian cut?

We believe this is possible, *and* is highly desirable, because statistical mechanics is difficult to understand and apply. We shall present a resolution where Planck's statistical mechanics is replaced by deterministic mechanics—viewing physics as a form of *analog computation with finite precision* with a certain dissipative

300

diffusive effect, which we model by digital computational mechanics associated with a certain numerical dissipation.

It is natural to model finite precision computation as a dissipative/diffusive effect, since finite precision means that small details are lost as in smoothing by the damping of high frequencies—which is the effect of dissipation by diffusion.

We consider computational mechanics in the form of the *General Galerkin (G2)* method for the wave equation, where the dissipative mechanism arises from a weighted least-squares residual stabilization[17]. We shall first consider a simplified form of G2 with least-squares stabilization of one of the residual terms and corresponding simplified diffusion model. We then comment on full G2 residual stabilization.

Wave Equation with Radiation and Dissipation

We consider the wave equation (4) with radiation augmented by (simplified) G2 diffusion:

$$\ddot{u} - u'' - \gamma \ddot{u} - \delta^2 \dot{u}'' = f, -\infty < x, t < \infty \quad (15)$$
$$\dot{E} = \int f\dot{u}\, dx - \int \gamma \ddot{u}^2\, dx, -\infty < t < \infty$$

where $-\delta^2 \dot{u}''$ models dissipation/diffusion from velocity gradients, $\delta = h/T$ represents a smallest coordination length with h a precision or smallest detectable change, and T is temperature related to the internal energy E by $T = \sqrt{E}$.

The relation $\delta = \frac{h}{T}$ takes the form $|\dot{u}|\delta \sim h$ with $T \sim |\dot{u}|$. A signal with $|\dot{u}|\delta < h$ cannot be represented in coherent form and thus cannot be emitted. This is like the 'Mexican Wave'

---

[17] J. Hoffman and C. Johnson, *Computation Turbulent Incompressible Flow*, Springer 2008.

around a stadium—which cannot be sustained unless people raise their arms properly; the smaller the 'lift' (with lift as temperature), the longer is the required coordination length or wavelength.

We see that the wave equation is here augmented by an equation for the internal energy E, which thus has a contribution from the dissipation $\int \delta^2 (\dot{u}')^2 dx$ (obtained as above by multiplication by $\dot{u}$).

In particular we have as above if the incoming wave is an emitted wave $f = -\gamma \ddot{U}$ of amplitude U, then:

$$\dot{E} = \int \gamma \left(\ddot{U}\ddot{u} - \ddot{u}^2 \right) dx \leq \frac{1}{2} \int \gamma \left(\ddot{U}^2 - \ddot{u}^2 \right) dx. \quad (16)$$

We assume incoming frequencies are bounded by a certain maximal frequency ν_{max}, we choose $\gamma = \nu_{max}^{-2}$ and assume $\nu_{max}^{-1} \gg \delta^2 = \nu_{cut}^{-2} \gg \gamma$, where $\nu_{cut} < \nu_{max}$ is a certain cut-off frequency.

We motivate this setup as follows: If u is a wave of frequency ν in x, then for $\nu > \nu_{cut} = \dfrac{T}{h} = \dfrac{1}{\delta}$, we have

$$\delta^2 \ddot{u}'' \sim \frac{h^2 \nu^2}{T^2} \dot{u}$$

which signifies the presence of considerable damping in (15) from the dissipative term, since $\dfrac{h^2 \nu^2}{T^2} \geq 1$. Alternatively, we have by a spectral decomposition as above:

$$\delta^2 \nu^2 \dot{u}_\nu^2 \sim f_\nu^2$$

and thus since $\gamma \ll \delta^2$

$$R_v = \frac{\gamma}{\delta^2}\delta^2 v^2 \dot{u}_v^2 \ll f_v^2.$$

Thus absorbed waves with $v > v_{cut}$ are damped and not fully radiated. The corresponding missing energy contributes to the internal heat energy E and increasing temperature T.

We will also find a cut-off for lower frequencies due to the design of the dissipative term $\delta^2 \dot{u}''$ corresponding to a simplified form of G2 discretization. In real G2 computations, the cut-off will have little effect on frequencies smaller than v_{cut}. In our analysis, we assume this to be the case, which corresponds to allowing δ to depend on v so effectively $\delta = 0$ for $v \leq v_{cut} = \frac{1}{\delta}$. We then obtain a Planck Law of the form

$$R_v(T) = \gamma T v^2 \theta_h(v,T) = \gamma T min(v^2, v_{cut}^2) \quad (17)$$

with a computational high-frequency cut-off factor $\theta_h(T) = 1$ for $v \leq v_{cut}$ and $\theta_h(v,T) = \frac{v_{cut}^2}{v^2}$ for $v_{cut} < v < v_{max}$ with $v_{cut} = \frac{T}{h}$.

Clearly, it is possible to postulate different cut-off functions $\theta_h(v,T)$ for example, exponential cut-off functions with the effect that $\theta_h(v,T) \approx 0$ for $v \gg v_{cut}$. In the next section we study the cut-off in G2.

The net result is that absorbed frequencies above cut-off will heat the string, while absorbed frequencies below cut-off will be radiated without heating (in the ideal case with the dissipation only acting above cut-off).

If the incoming radiation has a Rayleigh-Jeans spectrum $\sim \gamma T v^2$, then so has the outgoing radiated spectrum $R_v(T_v) \sim T v^2$ with $T_v \sim T$ for $v \leq v_{cut}$. In particular, the outgoing radiated spectrum is equilibrated with all colors having the same temperature, if the incoming spectrum is equilibrated.

Claes Johnson

Another way of expressing this fundamental property of the vibrating string model is to say that frequencies below cutoff will be absorbed and radiated as *coherent* waves, while frequencies above cut-off will be absorbed—transformed into internal energy in the form of incoherent waves, which are not radiated. Thus, high frequencies may heat the body and decrease the coordination length and allow absorption and emission of higher frequencies.

Note that the internal energy E is the sum over the internal energies E_ν of frequencies $\nu \leq \nu_{cut} \sim T$ with $E_\nu \sim T$ assuming equilibration in temperature, and thus $E \sim T^2$ motivating the relation $T = \sqrt{E}$.

Cut-Off by Residual Stabilization

The discretization in G2 is accomplished by residual stabilization of a Galerkin variational method and may take the form: find $u \in V_h$ such that for all $v \in V_h$

$$\int (A(u) - f)v \, dxdt + \delta^2 \int (A(u) - f)A(V)dxdt = 0,$$
(18)

where $A(u) = \ddot{u} - u'' - \gamma\dddot{u}$ and V is a primitive function to v (with $\dot{V} = v$), and V_h is a space-time finite element space continuous in space and discontinuous in time over a sequence of discrete time levels.

Here $A(u) - f$ is the residual and the residual stabilization requires $\delta^2(A(u) - f)^2$ to be bounded, which should be compared with the dissipation $\delta\ddot{u}^2$ in the analysis with $\ddot{u}^2$ being one of the terms in the expression $(A(u) - f)^2$. Full residual stabilization has little effect below cut-off, and acts like simplified stabilization above cut-off, and effectively introduces cut-off to zero for $v \geq v_{max}$ since then $\gamma|\ddot{u}| \sim \gamma v^2|\dot{u}| = \frac{v^2}{v_{max}^2}|\dot{u}| \geq |\dot{u}|$ which signifies massive dissipation.

The Sun and the Earth

If an incoming spectrum of temperature T_{in} is attenuated by a factor $\kappa \ll 1$ (representing a solid viewing angle $\ll 180°$), so that the incoming radiation $f_v^2 = \kappa\gamma T_{in}v^2$ with cut-off for $v > \frac{T_{in}}{h}$ (and not for $v > \frac{\kappa T_{in}}{h} \ll \frac{T_{in}}{h}$).

This may represent the incoming radiation from the Sun to the Earth with $\kappa \approx (\frac{R}{D})^2 \approx 0.005^2$—the viewing angle of the Sun seen from the Earth, where R is the radius of the Sun and D the distance from the Sun to the Earth. The amplitude of the incoming radiation is thus reduced by the factor κ, while the cut-off of the spectrum is still $\frac{T_{in}}{h}$.

The Earth at temperature T acts like a vibrating string and will convert absorbed radiation into heat for frequencies $> \frac{T}{h}$, that is, as long as $T < T_{in}$, while radiating $\sim \gamma T^4$ and absorbing $\sim \gamma\kappa T_{in}^4$ thus reaching equilibrium with $\frac{T^4}{T_{in}^4} \approx \kappa$. With $T_{in} = $ 5778K and $\kappa = 0.005^2$ this gives $T \approx 273K$ (including a factor 4 from the fact that the disc area of the Sun is πR^2 and the Earth surface area $4\pi r^2$ with r the Earth's diameter).

The amplitude of the radiation/light emitted from the surface of the Sun at 5778 K, when viewed from the Earth, is scaled by the viewing angle (scaling with the square of distance from the Sun to the Earth), while the light spectrum covering the visible spectrum centered at 0.5μm remains the same. The Earth emits infrared radiation (outside the visible spectrum) at an effective

Claes Johnson

blackbody temperature of 255 K (at a height of 5 km), thus with almost no overlap with the incoming sunlight spectrum. The Earth absorbs high-frequency reduced-amplitude radiation and emits low-frequency radiation, and acts as a transformer of radiation from high to low frequency; coherent high-frequency radiation is absorbed and dissipated in incoherent heat energy, which is then emitted as coherent low-frequency radiation.

The transformation only acts from high-frequency to low-frequency, and is an irreversible process representing the Second Law.

The Temperature of Radiation

The temperature T_{in} of incoming radiation with an attenuated Planck spectrum:

$$Rv = \kappa\gamma T_{in}v^2$$

with cut-off for:

$$v > \frac{T_{in}}{h},$$

can be read from the cut-off (Wien's Law), while the amplitude does not carry this information unless the attenuation factor κ is known.

For the outgoing spectrum $\kappa\gamma T_{in}v^2$, we noted that:

$$T \leq T_{in}$$

since heating requires dissipative cut-off after absorption, which requires that incoming radiation to contain higher frequencies than outgoing and that is only possible if the temperature of the incoming radiation is greater than the present temperature of the

306

absorbing body, as also expressed in the basic energy balance[18]: energy is transferred only from warmer to cooler.

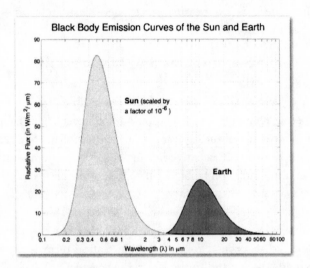

Figure 3: Blackbody spectrum of the Sun and the Earth.

A Fourier Law of Radiative Heat Transfer

Suppose an incoming radiation, with a spectrum of $\kappa\gamma Tin\nu$ and temperature Tin (with $\kappa \leq 1$), is absorbed and then emitted with spectrum $\gamma T\nu^2$. The heating effect from frequencies above cut-off T, assuming $h = 1$, is then given by:

$$\int_T^{Tin} \kappa\gamma T_{in}\nu^2 d\nu \sim \kappa\gamma T_{in}(T_{in}^3 - T^3) \sim \kappa\gamma T_{in}^3(T_{in} - T) \quad (19)$$

[18] J. Hoffman and C. Johnson, *Computation Turbulent Incompressible Flow*, Springer 2008.

which can be viewed as a Fourier Law with heating proportional to the temperature difference $T_{in} - T \geq 0$. Note: if $T_{in} < T$, then there is no heating since there is no cut-off—all of absorbed radiation is emitted.

The Second Law and Irreversibility

Radiative heating of a blackbody is an irreversible process, because the heating results from dissipation with coherent high frequency energy above cut-off being transformed into internal heat energy. We have shown that radiative heating requires the temperature of the incoming radiation be higher than that of the absorbing body.

We assume dissipation is only active above cutoff, while radiation is active over the whole spectrum. Below cut-off, radiation is a reversible process since the same spectrum is emitted and absorbed. Formally, the radiation term is dissipative and would be expected to transform the spectrum. The fact that it does not is a remarkable effect of the resonance.

Aspects of Radiative Heat Transfer

We find aspects of radiative heating in many different settings, as heat conduction or communicating vessels—with the flow always from higher level (temperature) to lower level. But radiative heat transfer is richer in the sense that it involves propagation of both waves and energy.

Let us try with a parallel in psychology. We know trivial messages radiated from a parent may enter one ear of a child and exit through the other, while less trivial messages would be completely ignored. However, the alertness of the child may be raised as a result of a 'high temperature' outburst by the parent opening the child's mind to absorbing/radiating less-trivial messages.

Here, we distinguish between the propagation of a message and its meaning.

Reflection vs. Blackbody Absorption/Emission

A blackbody emits what it absorbs $(f^2 \rightarrow R)$, and it is natural to ask what makes this process different from simple reaction (e.g. $f \rightarrow -f$ with $f^2 \rightarrow f^2$)? The answer is that the mathematics and physics of blackbody radiation $f \rightarrow \ddot{u} - u'' - \gamma\dddot{u}$, is fundamentally different from simple reflection $f \rightarrow -f$.

The string representing a blackbody is brought to vibration in resonance with the forcing and the vibrating string emits resonant radiation. Incoming waves are absorbed into the blackbody (string) and then are emitted, depending on the body temperature. In simple reflection there is no absorbing/emitting body, just a reflective surface without temperature.

Blackbody as Transformer of Radiation

The Earth absorbs incident radiation from the Sun with a Planck frequency distribution characteristic of the Sun's surface temperature of about 5778 K and an amplitude depending on the ratio of the Sun's diameter to the distance of the Earth from the Sun. The Earth, as a blackbody, transforms the incoming radiation to an outgoing blackbody radiation of temperature about 288 K, so the total incoming and outgoing energy balances.

The Earth acts as a transformer of radiation and transforms incoming high-frequency low-amplitude radiation to outgoing low-frequency high-amplitude radiation under conservation of energy.

This means high-frequency incoming radiation is transformed into heat which presents itself as low-frequency outgoing infrared radiation, so that the Earth emits more infrared radiation than it absorbs from the Sun. This increase of outgoing infrared radiation is not an effect of backradiation, since it would also be present if there were no atmosphere.

The spectra of incoming blackbody radiation from the Sun and outgoing infrared blackbody radiation from the Earth have little overlap, which means the Earth, as a blackbody transformer, distributes incoming high-frequency energy so that all frequencies below cut-off obtain the same temperature. This connects to the basic assumption of statistical mechanics of *equidistribution in energy* or thermal equilibrium with one common temperature.

In the model above, the absorbing blackbody inherits the equidistribution of the incoming radiation (below cut-off) and emits an equidistributed spectrum. To ensure an emitted spectrum is equidistributed, even if the forcing is not, requires a mechanism driving the system towards equidistribution (thermal equilibrium).

Connection to Turbulence

The computational dissipation in our radiative model acts like turbulent dissipation in slightly viscous flow, in which high-frequency coherent kinetic energy is transformed into heat energy in the form of small-scale incoherent kinetic energy. The small coefficient γ in radiation corresponds to a small viscosity coefficient in fluid flow.

Since γ is small, the emitted wave is in one sense a small perturbation, but this is compensated by the third order derivative in the radiation term, with the effect that the radiated energy is not small. Or, expressed differently: temperature involves first derivatives (squared) and radiated energy a second derivative multiplied by a small factor. Without the dissipative radiation term, the string cannot emit the energy absorbed and the temperature will increase without limit. With radiation, the

310

temperature will be limited by the temperature of the incoming wave.

Climate Alarmism and Backradiation

*It is virtually certain that increasing atmospheric con-
centrations of carbon dioxide and other greenhouse gases will
cause global surface climate to be warmer.*
—American Geophysical Union

*We know the science, we see the threat, and we know the
time for action is now.*
—Arnold Schwarzenegger

*There are many who still do not believe that global warming
is a problem at all. And it's no wonder: because they are the
targets of a massive and well-organized campaign of
disinformation lavishly funded by polluters who are
determined to prevent any action to reduce the greenhouse
gas emissions that cause global warming out of a fear that
their profits might be affected if they had to stop dumping so
much pollution into the atmosphere.*
—Al Gore

Global climate can be described as a thermodynamic system with gravitation subject to radiative forcing by blackbody radiation. Understanding climate requires understanding blackbody radiation. A main lesson of this note is that 'backradiation' is unphysical because it is unstable and serves no role, and thus should be removed from climate science, cf. Figure 4.

Since climate alarmism feeds on a 'greeenhouse effect' based on 'backradiation', removing backradiation removes the main energy source of climate alarmism.

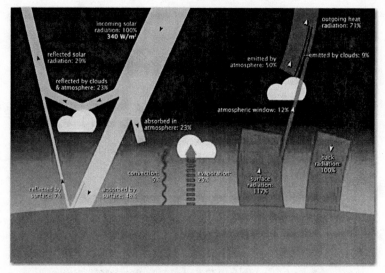

Figure 4: The Earth's Energy Budget According to NASA[19]

---

[i] *I consider it quite possible that physics cannot be based on the field concept, i.e., on continuous structures. In that case, nothing remains of my entire castle in the air, gravitation theory included, and of the rest of physics.*
—Albert Einstein (1954)

[ii] *What I wanted to say was just this: in the present circumstances the only profession I would choose would be one where earning a living had nothing to do with the search for knowledge.*
—Albert Einstein's last letter to Max Born Jan 17, 1955 shortly before his death on the 18th of April, probably referring to Born's statistical interpretation of quantum mechanics.

---

[19] Note the incorrect unphysical 100% backradiation and 117% = $390W/m^2$ outgoing radiation from the Earth's surface, but with correct physical 30% out of absorbed 48% transported by convection/evaporation from the Earth surface to the atmosphere.

Chapter 20

Do IR-Absorbing Gases Warm or Cool the Earth's Surface?

by Charles R. Anderson

THE TITLE OF this chapter is no doubt jolting to most readers. It is always assumed—by the catastrophic man-made global warming advocates—that infrared-absorbing water vapor, carbon dioxide, and methane gas, called greenhouse gases, are responsible for greatly warming the surface of the Earth.

In fact, as seen from space, the Earth has a 'blackbody' radiation temperature of about 255 Kelvin or 255K for short. The Earth's surface commonly has an average temperature of about 288K. The difference in these temperatures of about 33K or 33°C is attributed to the so-called greenhouse gas effect by most such advocates.

This implies a large warming effect. The articles called *The Earth's Gravitational Field*[1] and *Near Sea Level Atmospheric Temperatures*[2] shows that a part of this 33K temperature difference is not created by infrared (IR) absorbing gases, but instead is generally a result of the Earth's gravitational field acting on the gases of our atmosphere. This chapter will show that most of this

---

[1]

[2] —Charles R. Anderson. Available by request? Link?

313

33K temperature difference is accounted for by the 'blackbody' radiation balance with the spherical envelop of the atmosphere in radiative equilibrium with space. We will note the huge heat capacity of the oceans, the land surface, and the atmosphere contributes additional warming effects. There is no need to resort to a nonsensical greenhouse gas theory to describe a warming effect accounting for the supposed 33K problem.

The hypothesis that IR-absorbing gases are responsible for the large 33°C temperature difference between the Earth's measured 'blackbody' temperature of 255K and the average sea level surface temperature of about 288K has large obstacles to overcome.

Proponents of the greenhouse gas warming hypothesis claim solar radiation is transmitted through our atmosphere in the short wavelength portions of the electromagnetic spectrum as ultraviolet, visible light, and relatively short wavelength infrared radiation. This radiation is absorbed by the surface of the Earth and warms it. The surface then emits long wavelength infrared radiation upward into the atmosphere. They say the infrared-absorbing gases in the atmosphere absorb the IR radiation and re-emit half of it into space and half of it back toward the surface of the Earth.

They claim the half re-emitted back to the Earth's surface is absorbed by the surface and then re-emitted toward the atmosphere. A second time, IR-absorbing gases absorb this IR radiation and half of the half is emitted again toward the Earth's surface. This process repeats infinitely and the net result of adding up all the halves of halves of halves, etc., is a doubling of the warming power of the solar radiation initially incident upon the surface. This interesting theory violates the conservation of energy law—stated plainly, it does not happen.

Let us examine some properties of blackbody radiation for a moment. The relations between power in Watts (W) radiated by

Do IR-Absorbing Gases Warm or Cool the Earth's Surface?

a blackbody sphere at a temperature T given in Kelvin is described by the Stefan-Boltzmann Law (SBL) formula:

$$P = A\sigma T^4$$

Where P is total emitted radiant heat energy
A is the area in square meters
σ = blackbody constant $5.6697 \times 10 \; W/m^2 K^4$
The area of a sphere of radius r is $4\pi r^2$

Now, as discussed in *The Earth's Gravitational Field* and *Near Sea Level Atmospheric Temperatures*, at the altitude of 5,000 meters above sea level, the temperature of the U.S. Standard Atmosphere of 1976 is 255K, which is the Earth's blackbody radiation temperature as seen from space.

The Earth's radius is about 6,376,000 meters, so the sphere in radiant equilibrium with space has a radius slightly larger—about 6,376,000 meters. If this sphere's surface were uniformly at the temperature of 255K, then its total radiant outward power would be 1.225×10^{17}W.

That sphere would also emit a total inward radiant power of the same amount and all inside wall areas of the sphere would be in equilibrium. It does not make a bit of difference whether this sphere is filled with greenhouse gases or not, provided there are no other sources of energy and no other mechanisms to dissipate power.

If we assume the sphere with the temperature of 255K is in equilibrium with a slightly smaller blackbody sphere with the radius of the Earth at sea level, we can calculate the temperature of that surface given that it must radiate a power equal to the power of the surrounding sphere which is in equilibrium with space.

The temperature will be higher, since the surface area of the sphere is smaller.

In fact, the temperature of the Earth's surface as a blackbody would be 255.100K or 0.1°C warmer than the sphere at an altitude of 5,000 meters above sea level where we find equilibrium with space.

But the Earth's surface is not really a blackbody, so the Stefan-Boltzmann equation must include an emissivity multiplication factor on the temperature side of the equation.

For the Earth's surface, this emissivity factor is about 0.7. In equilibrium, this requires the Earth's surface to be at an elevated temperature of 278.89K. This is only about 9K or 9°C below its average temperature of 288K. Anything otherwise violates the Energy Conservation Law.

Greenhouse gases cannot change this result, unless they are a source of energy, which they are not. All IR-resonant gases can do is capture energy for an instant and then release it, either by radiating it away or by collision with another gas molecule such as a nitrogen or an oxygen molecule—thus transferring heat energy to them. They then may radiate energy or transfer more of it through convection and gas collisions. But, none of these effects do more than transfer energy—i.e., move it around.

IR-resonant gases do not create energy.

Another basic reason the greenhouse gas or IR-absorbing gas theory of emitted, half absorbed, and re-emitted, then half absorbed scheme does not work is because photons of radiation inside a blackbody radiator do not behave like ordinary particles. Radiation from the walls of the blackbody vary to keep the hollow interior of a constant-temperature sphere in equilibrium.

The volume-energy density remains constant, even if you expand the sphere and make it bigger. To maintain equilibrium of a constant volume-energy density, the sphere's walls produce more photons. This is non-intuitive for most people. Indeed, it is non-intuitive to many people who study physics. For example, atoms cannot be spontaneously created like photons.

This causes problems with thinking you can follow the emissions of individual photons—count them and figure out how many are absorbed by IR-absorbing gases and then how many photons are emitted by the excited gas as radiation versus how much of the energy absorbed by the IR-absorbing gas is lost due to collisions with the many other gas molecules in the lower atmosphere. This is a real problem, since below about 4,000 meters altitude, more energy is transferred by collisions with nitrogen and oxygen molecules, than is transferred by radiation.

To further complicate things, energy is absorbed-emitted by the evaporation-condensation of water and the freezing-melting-sublimation of ice *and* moved around and by convection currents of air. These energy transfer mechanisms are the reasons why the Earth's surface itself is not in thermal radiative equilibrium with space while the Top Of Atmosphere (TOA) sphere at 5,000 meters in altitude...is.

If it were, the Earth's average surface temperature would be 278.89K as we calculated above. The fact that the surface is about 288K instead, tells us that IR-radiation is not the only reason the surface of the Earth is so warm averaged over a period of days. In this chapter, we will search for the mechanism that causes this additional 9K increase of surface temperature.

The scheme of following energy carried off the Earth's surface by IR emission is unmanageable and makes no sense. While problems with the naive IR-absorbing gas hypothesis are not immediately obvious to many, there is little excuse for the failure to understand this long before many tens of billions of dollars had been wasted on greenhouse gas research.

Interestingly enough, the concept of creating photons to maintain the interior volume of a hollow blackbody shell with a constant energy density as the sphere expands was discussed in my sophomore-year thermodynamics textbook[3].

Of course, the sphere around the Earth with a radius 5,000 meters greater than that of sea level is not really at a constant temperature, since part of the Earth is in daylight and part is in nighttime. Nonetheless, the above calculation gives a good sense of the magnitude of real radiant effects by blackbody radiators.

It makes it very clear: any real effects of IR-absorbing gases— are not of the scale of 33°C.

But, there are issues of interest that remain to be examined. One important one is that, often, the Earth's surface is not in equilibrium with the TOA sphere about 5,000 meters above it. The ground or the surfaces of the oceans, with massive heat capacities, retain heat obtained during the daytime for release in the night. Also, the temperature at the surface and at an altitude of 5,000 meters is certainly a function of how much of the solar radiation reaches deep into our atmosphere—into the lower few thousand meters and to sea level. If the atmosphere absorbed more of the UV, visible, and IR spectrum of the incoming solar radiation, the result would be: a cooling of the Earth's surface. More of the sun's heat might be retained in the upper atmosphere.

A very interesting article by Martin Hertzberg, Hans Schreuder, and Alan Siddons called *A Greenhouse Effect on the Moon?*[4] will be summarized here and discussed in this context.

The moon has no atmosphere and it is the same distance from the sun as the Earth is. Yet, the mid-day temperature on the moon's surface is about 370K or 97°C, which is about 20K cooler than expected just due to the radiation incident from the moon.

---

[3] *Thermal Physics,* Philip M. Morse, W.A. Benjamin, Inc., 1965
[4]

http://www.ilovemycarbondioxide.com/pdf/Greenhouse_Effect_on_the_Moon.pdf

Do IR-Absorbing Gases Warm or Cool the Earth's Surface?

The nighttime temperature gets down to about 85K or -188°C, but this is about 60K warmer than the expected low temperature.

Here's the reason: the surface of the moon has heat storage capacity...the subsurface remains somewhat cooler than the surface during its day when solar energy absorbed and it releases heat energy during its night. This makes the average temperature of the moon's surface is about 40K cooler than it would be otherwise. Analogously, the Earth's land surface, oceans (covering 70% of the surface) and atmosphere all have a heat capacity and modulate the flow of heat energy from the interior to the surface.

The heat capacities of the Earth's land and water and atmosphere exceed that of the rock of the moon, so the day-to-night moderating effect observed on the Earth is much larger than the moon's. This may well be the source of the additional 9K temperature increase found at the Earth's surface.

What is the effect of tiny proportions of IR-absorbing gases in our atmosphere when compared to the heat capacity and heat-limiting diffusion effects of the Earth's surface?

When discussing any theorized effects of IR-absorbing gases, one must account for the absorption of incident solar IR radiation in the Earth's atmosphere, which is very cavalierly disregarded by strong greenhouse gas-effect advocates who prefer offering a back-reflection (or backradiation) mechanism for warming. This is important because much of the sun's IR radiation reaches the Earth's surface and warms it directly, but some is absorbed in the atmosphere *before* it reaches the surface.

In addition, some of the sun's IR radiation is reflected by the surface instead of being absorbed, so it does not directly warm the air or the surface.

So, a key question arises: do atmospheric IR-absorbing gases create a net warming or net cooling effect on the Earth's surface?

First of all, let's enlarge the context of the discussion. The primary source of heat for the surface of the Earth is the radiant energy of the sun (called insolation). The solar wind of the sun, materials dumped into the atmosphere from space, heat from the deep interior of the earth, the interplay of changes in the Earth's magnetic field and the sun's magnetic field, and energy from the tidal effects of the gravitational interaction with the moon also contribute heat, though the sum of these is generally considered to be much less than that from the sun's radiant energy spectrum of ultraviolet (UV), visible and infrared (IR) light. The common explanation for the catastrophic greenhouse gas hypothesis ignores the effects of the incident IR portion of the sun's emission spectrum.

This is foolish.

UV light is 11% of the radiant energy from the sun. The solar cycle variation of UV light, at 0.5 to 0.8%, is much larger than the visible light variance (0.22%). UV light is absorbed throughout the atmosphere, but much still reaches the ground and is absorbed there. The amount of UV radiation absorbed in the upper atmosphere is dependent upon the amount of ozone. The amount of ozone is said variously to depend on the solar wind, chlorofluorocarbon compounds (CFCs), water vapor, and volcanic activity. When more UV light is absorbed in the stratosphere and does not reach the ground, its surface warming effect is greatly diminished.

The absorbed energy is re-emitted as IR radiation and much of that energy is quickly lost to space. Nonetheless, much of the UV light energy is absorbed by the ground.

It is often—incorrectly—said that the entire atmosphere is transparent to visible light—which is 40% of the sun's radiant energy. Visible light is reflected from clouds and aerosol particles, but, as we will see below, a considerable fraction of the visible light radiation does not reach the ground or oceans to warm their surfaces even when the sky is clear.

Finally, IR radiation is not strongly absorbed by the nitrogen, oxygen, and argon gases which make up 99% of the atmosphere. So, a large fraction of the solar IR directly warms the Earth's surface. Some is absorbed by the dominant IR-absorbing gas, water vapor, and small amounts are absorbed by oxygen, nitrogen, carbon dioxide and other IR-absorbing gases. The incoming IR radiation absorbed in the atmosphere is less effective in warming the Earth's surface than is that which is absorbed by the Earth's surface directly.

This is because much of the absorbed energy locally warms a mass of air which then expands, becoming less dense and more buoyant, and rises. Some of this energy absorbed in the atmosphere is re-radiated as IR radiation, but now half or more is directed outward into space.

In other words, with respect to incoming IR energy from the sun, more water vapor and CO_2 in the atmosphere results in less-effective surface warming. Thus, IR-absorbing gases have a cooling effect on the ground. In the radiance spectrum, 49% of the solar energy is in the IR wavelengths. This energy is still deposited in the Earth's atmosphere, but has less ability to warm the Earth's surface.

The solar light spectrum outside the atmosphere and the spectrum transmitted through the atmosphere to sea level are shown in Figure 1.

The measurement of the transmitted energy—and its distribution with wavelength—is highly dependent upon the amount of water vapor in the atmosphere, so the transmitted spectrum may vary considerably. But, for the purposes of this discussion, let's use the overall transmittance values to the Earth's surface from Figure 1, a graph of an actual particular measurement. The overall energy transmittance is about 0.65. The transmittance of UV and visible radiation combined is about 0.59, while that for IR radiation is about 0.69.

In each case—whether UV, visible light, or IR—not all of the radiation that strikes the Earth's surface is absorbed. Some is reflected and the amount of reflection depends on whether the ground is covered with snow, is plowed earth, covered with

grasses, forests, crops, blacktop, or water. There *are* some real
ways man effects the Earth's temperature. We change the surface
of the earth where we live...on a fraction of the 30% of the
Earth's surface which is land. We convert fossil and biomass fuels
into heat. We release carbon black, other small particles and
aerosols into the atmosphere, which have some impact on the
temperature at the Earth's surface.

Compared to natural variations, these man-made effects are
tiny, yet they are large compared to the effect of our adding CO_2
and methane to the atmosphere.

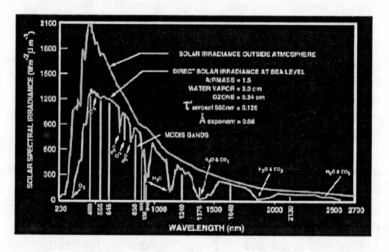

Figure 1: Solar Radiance and Emission Wavelength

Wherever the atmosphere is heated, there heat is transferred. In
the outer, very low-density parts of our atmosphere, the primary
means of heat transfer is radiant transfer by IR emission from an
energetic molecule or atom, since collisions of molecules and
atoms for direct energy transfer are rare. In the denser, lower-
altitude atmosphere, most energy transfer is due to gas molecule
collisions and the convective flow of masses of warmed air.

Do IR-Absorbing Gases Warm or Cool the Earth's Surface?

Near the Earth's surface, most of the energy lost by the warmed surface is due to gas molecules striking the surface and picking up heat and then colliding with other molecules to transfer heat from one to another.

Once a body of air is so heated, masses of warmed air molecules are transported upward into the cooler atmosphere at higher altitudes or laterally toward cooler surface areas by convection. Warmed molecules, most of which are nitrogen, oxygen, and argon, radiate IR radiation.

However, low-temperature molecules or atoms near the Earth's surface are ineffective energy radiators, as shown by the Stephan-Boltzmann equation where radiation is dependent upon the fourth power of the absolute temperature.

Thus, gas molecule collisions, and evaporation-condensation-freezing-melting of water and its convective transport are the dominant means of heat transfer. These processes, on balance, cool the surface of the Earth and redistribute some of the heat into the upper atmosphere and cooler places such as those shaded from the sun or the arctic regions.

Just outside the Earth's atmosphere, solar irradiance has a power density of about 1367 W/m^2. We saw from the discussion of the transmittance spectrum of the sun's radiation that the overall energy reaching the surface is about 65% of the total energy outside the outer atmosphere. So 0.65 times 1367 W/m^2 is 889 W/m^2, the amount that reaches the Earth's surface. Of this energy, reflected energy is equal to 1 minus the emissivity, whose average value is about 0.7 for the Earth's surface, so about 30% is reflected from the Earth's surface without being absorbed.

Thus, the energy warming the surface is about 622 W/m^2. When the Earth's surface temperature during the day at full sunlight is 290K or about 17° C and assuming the surface emissivity of the Earth is 0.7, the IR radiation of the surface is about 281 W/m^2. Thus, radiative cooling of the surface removes about 281 W/m^2 during the full sunlight day at a surface temperature of 290K.

The fraction of the cooling of the surface due to radiative cooling, r, is then about 281/622 or 0.45 during full light. This fraction is taken as 80% in some alarmist greenhouse warming calculations. The remaining cooling is by direct contact of the air with the surface, by evaporative cooling, and by the subsequent movement of masses of air in convection currents carrying that energy further away from the warm surface areas.

Since the dominant source of energy warming the surface of the Earth is the sun, let us do a simple calculation based upon the facts presented above. We will perform the calculation for a time of day with full light. Let us say that greenhouse gases absorb a fraction f of the incoming IR radiation from the sun, which is 49% of the sun's incoming energy.

From Figure 1, f is about 0.31 for the IR portion of the spectrum. Thus the energy absorbed by IR-absorbing gases from the incoming spectrum of solar energy is $0.49f$ or 0.15 and a fraction of this, say k, is radiated back into space without coming near the surface. NASA says k is 0.5, but it is actually slightly larger than that given that much of this absorption occurs at appreciable altitudes where the mean free path for radiation absorption is long.

This means the constant altitude surface is not well-represented by a half-plane. The total cooling of the ground due to IR-absorbing gases intercepting IR radiation before it reaches the ground is now $0.49fk$, or here about 0.075. Of this energy, had it become incident upon the surface as IR radiation, a part would have been reflected rather than absorbed. The fraction of the incoming IR radiation that would have been absorbed at the surface rather than reflected is q.

The net energy lost to the warming of the surface is then $0.49fkq$, or here about $0.075q$. This energy may be viewed as a cooling of the surface caused by IR-absorbing gases in the atmosphere, because on average the captured radiation was captured further away from the surface than will be IR radiation being emitted from the surface and because any radiative cooling of the heated gas molecule results in radiation toward space.

The discussion that follows will be carried out similarly to that of the advocates of human-caused warming due to increases of the concentration of IR-absorbing gases in the atmosphere such as carbon dioxide. The basis for the argument is scientifically suspect because of the properties of radiation from warm bodies in the electromagnetic spectrum and the easy creation of photons, especially the very low-energy photons characteristic of the long-wavelength infrared spectrum emitted by bodies at low temperatures similar to those on the Earth's surface and in the lower atmosphere where the temperature is below 300K.

Also, the detailed properties of the emitting surface of the blackbody radiator, in terms of its excitation states and the frequencies of the emitted photons, are not important for the thermodynamics. The point of the following exercise is to consider some of the issues of making an argument in the fashion of the CO_2 warming advocates. We also want to get some feel for the scale of possible effects of greenhouse gases. We already know the scale of CO_2 effect is small, but after going through the argument, this assertion will be further justified.

We will also see that its effect is certainly a net cooling effect, not a net warming affect as is invariably claimed for it.

Now, let's suppose a fraction g of the total energy from the sun reaches the Earth's surface. For the case of the transmittance graph above, g is about 0.65. Of the energy g absorbed in the surface, only r times it is emitted as IR radiation and that value was estimated above to be about 0.45.

Since the IR-absorbing gas content of the atmosphere is unchanged, the amount of ground cooling IR radiation absorbed by IR-absorbing gases in the lower atmosphere, is now rgf'', where f'' is the fraction of the ground-emitted IR absorbed by the atmosphere. Because the distribution of the IR radiation wavelengths is different from the ground than from the sun, the previous f and the present f', are not the same thing.

Let us examine some data from which we can estimate the fraction f' in Figure 2.

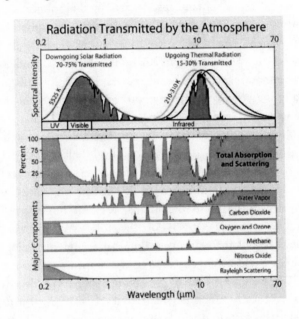

Figure 2: Radiation Transmitted by the Atmosphere

Note the solar radiation absorption spectrum at the top right of Figure 2 shows somewhat less absorption than the actual measurement in the Figure 1 of this article.

This probably reveals there is some shortcoming in the approach of trying to reconstruct that absorption from the separate absorption spectra of the gases considered here. Nitrogen gas, which is 78.084% of the atmosphere, is entirely left out. But since this data is well-respected in catastrophic greenhouse gas advocate circles, it is fair to use it to at least show some of the limitations of the usual explanations of the catastrophic greenhouse gas hypothesis.

The fraction of the long wavelength IR emitted from the ground at about 290K which is absorbed as actually shown in this figure is 0.65, though the labeling says it is from 0.70 to 0.85. Thus, we will take f'' to be 0.65.

A fraction, r'', of the gas molecules which absorb long wavelength IR radiation emitted from the ground will cool by emitting IR radiation in turn. Water vapor is the best long wavelength IR absorber and it is the best emitter of IR energy, but before it can emit the energy it absorbed from IR radiation, it's more likely to suffer numerous gas collisions where much of its excess molecular energy is transferred in those collisions to the molecules which collide with the water molecule.

Nitrogen molecules are the most likely to take up energy from the water molecule, since it's so prevalent in the atmosphere. Oxygen molecules are the next most-likely colliders at 20.95% and then argon atoms at 0.93%. Together, these three gases account for 99.97% of the U.S. Standard Atmosphere. None of these gas molecules are very efficient IR emitters in the long wavelength spectrum.

At sea level, the mean gas velocity is 459 m/s, the mean free path or distance between collisions is only 6.6×10^{-8} m or 66 nm, and the collision frequency is 6.9 billionths of a second.

At an altitude of about 4,000 m, the radiative transfer of energy competes about evenly with transfer by collisions. At that altitude, the time between gas molecule collisions is about 4.4 billionths of a second.

If we treat this very approximately as a means to estimate the time for the radiative transfer of energy from an excited state in a molecule, we may say the effective time is about 0.455 billionths of a second. At sea level, there is a gas molecule collision every 0.145 billionths of a second.

This suggests there are about three gas molecule collisions at sea level for every emission of a photon upon de-excitation of an excited state in a gas molecule.

What is more, in some wavelength zones in the electromagnetic spectrum of the infrared radiation, it would be likely that more than one absorption event of photons would have to occur in IR-absorbing molecules of the particular greenhouse gas such as CO_2 before energy radiated from the ground was lost to space at that wavelength. Before that could happen, the energy would likely be transferred to ordinary nitrogen, oxygen, and argon molecules or atoms by gas collisions.

This phenomenal number of collisions spreads the IR energy absorbed by a water molecule or a CO_2 molecule near the ground to the dominant nitrogen and oxygen molecules very, very quickly.

At an altitude of 5 km, the collision period is still 3.9 billionths of a second and at 10 km altitude it is 2.1 billionths. If a water molecule is to radiate energy away as IR emission, it must do so very quickly!

If it were able to emit IR very quickly, then the atmosphere would cool down very quickly at night. Indeed, cooling at high elevations in mountains—by radiative cooling—is more rapid than cooling from sea level because less of the radiative energy of the ground is spread to the many nitrogen and oxygen gas molecules that hold the energy.

The ground gives up approximately 45% of its energy by IR emissions and that energy would be absorbed by IR-absorbing gases with about 65% efficiency and half of that gas-absorbed energy would be quickly radiated off into space.

The half returned to the ground would soon be radiated again from the ground and the process would repeat. This would have to repeat on a time scale of billions of times a second to compete with gas collisions as the means of energy transport.

Do IR-Absorbing Gases Warm or Cool the Earth's Surface?

If it did, the atmosphere would cool at a catastrophic rate at night. It is a good thing that the long-wavelength highly-excitable IR-absorbing gases are not big players in the competition to remove energy from the Earth's surface at and near sea level.

Surface-emitted long-wavelength IR radiation energy is quickly spread from good IR absorbers to poor IR absorbers or emitters through billions of collisions per second. The energy is then transported through the atmosphere by particle collisions and warm convective currents which rise higher into the atmosphere. For the reasons discussed, it would be surprising if r'' is as large as 0.1, which I will use in this calculation for want of a better number at this time. A fraction, j, of this energy will be emitted by the IR-warmed greenhouse gas molecules back toward the ground. NASA says this fraction is 0.5.

Let's assume j is about 0.5. The greenhouse gas warming of the surface due to absorbing IR radiation from the ground and re-emitting it toward the ground would then be about $jrgr''f''q$, where q is the fraction of back-reflected IR radiation that was incident upon the surface and absorbed. Remember that some radiation is reflected. The reduction in IR cooling of the surface is then about $0.010q$.

There is another term for the IR radiation which is reflected from the surface without having been absorbed in the surface. The fraction of the incoming IR radiation reflected from the surface is $(1-q)$ and the fraction of the total incoming energy from the sun that was initially IR radiation was 0.49. The fraction of the IR radiation incident upon the surface is $(1-f)$.

The total of initial incoming solar radiation reflected from the surface is then $0.49(1-f)(1-q)$. Of this outgoing reflected IR radiation, a fraction h is absorbed by IR-absorbing gases; h is less than f, the fraction of the IR radiation absorbed by gases from initial incoming IR radiation from the sun. The reason h is smaller is because the IR radiation that made it through the atmosphere once was largely in frequency windows where little absorption

329

occurs. Examining the sea level solar spectrum in Figure 2, a reasonable approximation for h is about $0.5\,f$.

Once again, of the molecules absorbing IR radiation reflected from the surface, only a fraction r'' will re-emit IR radiation. Of the gas-absorbed IR radiation reflected from the surface, roughly half is re-emitted toward the surface and a fraction q of that is absorbed by the surface. The result is that this reflected IR contribution to warming the atmosphere closer to the surface is $0.245(1-f)(1-q)r''hq$. Using the value of f of 0.31 (from Figure 2) for the IR part of the solar spectrum, this term becomes $0.026(1-q)r''q$.

Now we will compare the greenhouse gas cooling effect on the incoming solar radiation of $0.45fkq$ to the decreased cooling of the surface due to IR ground emissions being absorbed by IR-absorbing gases and the reflected IR contribution of energy re-directed to the surface from IR-absorbing gases.

The value of r'' will be set at a conservatively high 0.45 to equal the radiative cooling fraction of the ground energy, even though the gas molecules will average a cooler temperature.

The ratio of the warming terms to the cooling term is:

$$(jrgr''f'q + 0.245(1-f)(1-q)r''hq) \,/\, 0.49fkq$$

$$= (0.010 + 0.012(1-q))/0.075$$

$$= 0.13 + 0.16(1-q)$$

Now, recall that q is the fraction of the solar IR incident at sea level which is absorbed by the Earth's surface. The term the greenhouse global warming alarmists carry on so much about is only about 13% of the cooling effect of IR-absorbing gases due to keeping heat away from the surface by absorbing the incoming solar radiation.

Do IR-Absorbing Gases Warm or Cool the Earth's Surface?

If all the ground-incident IR radiation is absorbed, q is 1 and the second term is zero. In the ridiculous case that q is zero, the sum of the two terms retarding the radiant cooling of the ground in ratio to the radiation which never warmed the surface is 0.29, and there is a substantial net cooling effect during the day due to the IR-absorbing gases. But, q is more likely to be about 0.7, in which case the ratio of the warming to the cooling is about 0.13 + 0.05 = 0.18. It seems clear that the addition of IR-absorbing gases to the atmosphere creates a net cooling effect during the period of daylight.

In comparison, the only effect at night is the backscatter of IR-radiation toward the ground as the ground is cooled by radiative cooling. Of course, gas collisions with the ground, evaporative cooling, and convective cooling also continue.

Since the ground cools, the radiative cooling will become slightly less effective as the temperature drops and T^4 becomes smaller.

The nighttime radiative cooling will be made less effective by the re-absorption of IR photons from IR-absorbing gases. However, there is another effect at night. The transport of energy by radiation is actually faster than by convection, so IR-absorbing gases can also have a competing cooling effect at night. Thus, during the day, the net effect of IR-absorbing gases is a cooling effect, while at night the net effect is more difficult to evaluate. It seems likely that it depends on the altitude and certainly on the amount of water vapor. There is so little CO_2, that I doubt its effect is measurable and indeed, no one actually does seem to have any relevant measurements.

This is why the climate models are so necessary to those who claim that CO_2 is an important warming gas. IR-absorbing gases play a role in moderating daytime temperatures, but beyond that the effect is not so clear. The heat capacity of our atmosphere, with its moderate radiative cooling, and the heat capacity of our oceans and the ground itself, each with slow heat diffusion, play a critical role in moderating swings in the daily temperature between night and day, between passing clouds and the reappearance of full sunlight, and over periods of several days, as well as around the year since the oceans tend to warm many land areas in the winters.

The cooling effect due to solar IR radiation absorption by IR-absorbing gases while the sun is shining is proportional to 1 - 0.18 = 0.82, though this value will be less very early or very late in the day.

The nighttime maximum decrease in cooling due to IR-absorbing gases is estimated to be about 0.13, though as mentioned, the speed of transmission of energy by radiation is then ignored and may have a countervailing effect.

Just comparing the size of the day number to the maximum night warming number, the net effect of IR-absorbing gases over the day is a cooling effect. But an hour-by-hour calculation of the absorbed solar IR during the day will average less than 0.82 as noted. The cooling effect will not be as large as the above numbers imply, but it is clear there is a net cooling effect, rather than a net warming effect, when averaged over the entire day.

Let us suppose one was to add more CO_2 to our atmosphere. CO_2 does not absorb across a wide range of wavelengths. Most of its long wavelength absorptive power occurs in a wavelength range in which water vapor is already strongly absorbing. Another strong absorption wavelength band (separate from the overlapping water absorption wavelength) is very narrow. The total absorption in both wavelength ranges is already very strong due to CO_2 at its present concentrations, so additions of CO_2 will have very little additional effect on any possible slowing of the nighttime cooling of the surface. The absorption effect is already virtually saturated.

The effect of additional CO_2 on the average temperature over a day due to slowing nighttime cooling will also be reduced by that addition bringing about additional daytime cooling by additional absorption of the incoming solar radiation well above the ground.

The usual claims of CO_2 warming are exaggerated and despite that, proponents of its catastrophic greenhouse warming power still have to conjure an effect-multiplier by postulating a positive feedback from additional water vapor. So, they suppose an increase in CO_2 both causes greater warming than it does and it causes a substantial increase in water vapor to cause a further increase in backscattered radiative warming.

If it were true that CO_2 increases lead to water vapor increases, then the daytime cooling effect of the increased water vapor would still cause a net cooling of the Earth's surface due to daytime additional absorption of sunlight by added water vapor and CO_2.

In sum, using simple calculations, we approximate the net effect of IR-absorbing gases on the surface temperature of the Earth. We conclude that the cooling effect during full sunlight hours caused by preventing incoming solar IR radiation from impacting the Earth's surface is several times greater than the possible retardation of the cooling effect due to IR-absorbing gases.

At night, IR-absorbing gases may retard radiative cooling of the ground, consistent with the observation that very humid nights cool less rapidly than very dry nights, though humidity acts to retard cooling also by making the air more dense and because water has a high heat content. Here, the non-radiative effects of water are more important. Now, if the effect were very large in either case, this might be cause for concern.

We would likely be better off heating the surface of the planet than cooling it and it is good that the IR-absorbing gas effect moderates temperature swings between night and day by cooling the days. Additions of IR-absorbing gases—whose absorbing effects are not now saturated—may principally serve to further moderate the differences in daytime and nighttime temperatures. This is an overall effect which is good for plants, animals, and humans.

Note one implication is that claims that additional IR-absorbing gases, such as CO_2, will cause additional melting of Arctic ice are unjustified. Arctic regions are so cold that any melting will occur mostly during daylight hours in summer months. Yet, that's the time when direct sunlight does much of the warming and, as we have seen, more incoming IR solar radiation will then be absorbed by IR-absorbing gases.

The net daytime cooling effect should mean: even if there were a slight average warming over the entire day, there would be a decrease in temperature during the hours melting mostly occurs. Thus, the net effect of IR-absorbing gases is a cooling effect. Hence, alarmist scenarios of massive amounts of Arctic ice melting due to increases in CO_2 are unfounded.

One cannot focus only on outgoing IR radiation due to light absorbed in the Earth's surface while ignoring the large part of the sun's total incident radiation which is IR from the get-go. One cannot ignore gas collisions, evaporative cooling, and convection currents as mechanisms for heat transfer from the ground to the atmosphere.

The fact that IR absorbed from the incoming solar spectrum occurs higher in the atmosphere on average and the energy there cannot be as effectively transported to the near sea level atmosphere or to the ground is very important.

But, gas molecular collisions, evaporative cooling, and convection can take the energy off the ground and transport it to higher altitudes to replace air cooled by radiating IR energy out in space. These processes are more suited to retaining heat near sea level than are the faster-acting radiative processes of heat transfer. This is very important.

The most important warming effect on the surface is radiation absorbed upon the incidence of direct solar radiation. During

daytime, additions of IR-absorbing gases result in more energy of
the solar spectrum being deposited in the atmosphere rather than
in the surface. This results in a net cooling effect and helps to keep
us from broiling at midday.

That hardly seems to hold the makings of catastrophe.

Chapter 21

Legal Fallout from False Climate Alarm
by John O'Sullivan

Now Governments Begin to Abandon Falsified National Temperature Records

WITH A DRAMATIC courtroom defeat in October 2010, the Sky Dragon, nurtured on the green religion of international climate alarmism, suffered a mortal legal wound. Only once before, in 2007 at the High Court in London, had that beast, so engorged on eco-propaganda, come so close to being slain. On that occasion the court ruled that former U.S. Vice President, Al Gore's film, *An Inconvenient Truth* contained nine critical factual errors and could no longer be shown in schools in England and Wales as a portrayal of fact. Bereft of a *bona fide* scientific underpinning that could be defended in court it became just a matter of time before astute lawyers ripped this chimera to shreds.

The Sky Dragon that blows hardest only in the dark recesses of ignorance and intellectual apathy had no puff to withstand a cutting battle in the legal light of reason. And this greatest of skeptic (climate realist) legal triumphs occurred not in the United States, where most analysts had expected, but in New Zealand. At the time of this book's publication (January, 2011) most observers

337

John O'Sullivan

had still not absorbed the stunning consequences of the Kiwi calamity; so few had understood that a pro-green western government had actually abandoned all pretense of possessing an 'official' climate record in what constitutes the most humiliating of climb downs.

Three years after that legal flesh wound was inflicted in the English High Court, New Zealand justice deals mortal injury upon that insane Sky Dragon. The New Zealand Climate Science Coalition (NZCSC) had demonstrated how certain western national governments' arguments for global warming are so easily shredded[1] when employing a long-standing legal tactic available to common law country citizens. I first drew attention to this strategy in my article *Prosecuting Climate Fraud: The International Dimension*[2].

The New Zealand government, now dragged into court, finally confessed that they and their fellow doomsayers could no longer refer to an official climate record—there isn't one[3]. A beast bloated on billions of tax dollars was left to stumble and fall onto a sharpened legal sword.

Earlier in the year, with the able assistance of my law associates around the globe, I explained how such a legal triumph could be won. In my online articles I showed bloggers the legal thread that not only linked the five English-speaking nations tied up in this great global warming swindle—the UK, US, Canada, Australia, New Zealand, but also provided the key to victory. All such Anglophone nations, while operating their independent legal systems, nonetheless premise themselves on English common law.

---

[1] NZCSC & Climate Science Conversation Group; Press Statement of December 18, 2009; accessed online (April 26, 2010);

[2] http://algorelied.com/?p=3768&cpage=1

[3] Bolt, A. 'Climategate: Making New Zealand warmer,' *Herald Sun* (November 26, 2009), accessed online April 26, 2010;

Under common law our respective governments cannot impose climate regulations on us by regarding similar facts and circumstances differently on different occasions. This principle is known among legal practitioners as *stare decisis* (i.e. judges are obliged to obey the set-up precedents established by prior decisions).

For example, we studied in detail two of ninety legal challenges filed in U.S. courts against the meritless federal climate legislation being brought in via the back door by the Environmental Protection Agency (EPA)[4].

It became clear that the EPA sought to impose upon the people 'arbitrary and capricious' governmental, climate-related decisions with little or no scientific justification.

All such challenges are traditionally referred to as *mandamus* petitions. My own use of this most valuable legal instrument has been the New York version of mandamus known as an 'Article 78' action. New Zealand's National Institute of Water and Atmospheric Research (NIWA) stood accused of repeatedly frustrating NZCSC in its attempts to get government climatologists to explain how they managed to create a warming trend for their nation's climate that is not borne out by the actual temperature record.

NZCSC petitioned the high court of New Zealand to force NIWA (effectively the Kiwi government) to validate their national weather service's reconstruction of antipodean temperatures—or strike it down. Ostensibly, NZCSC would present evidence in court that NIWA had faked their nation's climate data if they declined to disown it. The full petition may be read here[5].

---

[4] http://www.epa.gov/lawsregs/
[5] Dunleavy MBE, T.,'High Court asked to invalidate NIWA's official NZ temperature record,' (August 13,2010); climatescience.org.nz, (accessed online: October 6, 2010)

Before the matter could be put to the court for a final judgment NIWA's statement of defense gave up the fight. Their attorneys advised the court that NIWA never accepted responsibility for a national temperature record (referred to by them as the NZTR).

Thus by distancing itself from the indefensible NIWA confessed there was never any such thing as an 'official' NZ Temperature Record, despite there being an official government acronym for it (NZTR). Controverting all previous policy statements, the NZ government now wishes it to be known that the country has never maintained an official record; all such published data was only intended for internal research purposes and not as evidence to prove the country warmed due to human emissions of carbon dioxide.

However, all such data had shamelessly been hyped up via the IPCC as the gold standard of the entire New Zealand temperature history and for decades cited by pro-green advocates as proof of antipodean man-made climate warming. Along with the discredited Australian (BOM) records, the NZ numbers represented the cornerstone of Australasia/South Pacific (Oceania) warming. Significantly, this region constitutes two of the eight terrestrial ecozones; with such scant alternative records, we may now infer that at least one quarter of the world's 'official' climate record is discredited and an unjustified carbon tax is being extorted.

NZCSC had previously issued a joint press release with the Climate Science Conversation Group (December 18, 2009) accusing NIWA of publishing, 'misleading material'. The two organizations were rightly not letting up in their pursuit of access to government methods that were now so shrouded in secrecy. Repeated refusal to come clean led to charges that NIWA had been 'defensive and obstructive' in requests to see New Zealand climate scientists' data.

Downloadable pdf files of letters between Coalition chairman and barrister Barry Brill and NIWA chairman Chris Mace may be read here[6]. As we recall, the patterns here mirror Climategate; data is challenged for being dubious and then is either withdrawn or destroyed before tested in the courts. The evidence for global warming again melts away under the harsh light of courtroom scrutiny.

According to NZCSC, climate scientists cooked the books by using the same alleged 'trick' employed by British and American doomsaying scientists. This involves subtly imposing a warming bias during what is known as the *homogenization*[7] process that occurs when climate data needs to be adjusted. Indeed, the original Kiwi records show no warming during the 20th century, but after government sponsored climatologists had manipulated the data a warming trend of $1°C$ appeared.

Homogenization Explained

When such data adjustments (homogenizations) are made, scientists must keep their working calculations so that other scientists can test the reasonableness of those adjustments. According to an article in *Mathematical Geosciences*[8], homogenization of climate data needs to be done because 'non-climatic factors make data unrepresentative of the actual climate variation'. The great irony is that the justification made for the need to 'homogenize' data is because if it isn't then the

---

[6] Atkins, Holm, Joseph & Majurey., [Solicitors], *Statement of Defence on Behalf of the Defendant*, [On behalf of NIWA], (September 14, 2010);

[7] Costa, A.C. and A. Soares, *Homogenization of Climate Data: Review and New Perspectives Using Geostatistics*, Mathematical Geoscience, Volume 41, Number 3 / April, 2009;

[8] (April 2009)

'conclusions of climatic and hydrological studies are potentially biased'.

Did you get that? Climate scientists need to add their own spin to the raw temperatures because if they don't then they are less reliable!

However, according to the independent inquiry into Climategate chaired by Lord Oxburgh, it was found that it was the homogenization process itself that became flawed because climatologists were overly guided by 'subjective' bias. Notably, Australian Andrew Bolt, writing for *Herald Sun* sagely determined that the Kiwigate scandal was not so much about "hide the decline" but "ramp up the rise"[9]. Bolt goes on to report, "Those adjustments were made by New Zealand climate scientist Jim Salinger, a lead author for the Intergovernmental Panel on Climate Change (IPCC)". Salinger was dismissed by NIWA during 2010 for speaking without authorization to the media.

Pointedly, Salinger once worked at Britain's CRU (Climate Research Unit), the institution at the center of the Climategate scandal. Salinger became part of the inner circle of climate scientists whose leaked emails precipitated the original climate controversy in November 2009. In an email[10] to fellow disgraced American climate professor, Michael Mann, Salinger stated he was "extremely concerned about academic standards" among climate sceptics.

On January 29, 2010, in what seemed like a reprise of the Phil Jones debacle at Britain's CRU, the Kiwi government finally owned up that "NIWA does not hold copies of the original worksheets".

---

[9] Bolt, A. '*Climategate: Making New Zealand Warmer,*' Herald Sun (November 26, 2009), accessed online April 26, 2010;
[10] Salinger, J. Climategate email Filename: 1060002347.txt. (August 4, 2003).

Kiwigate Mimics Climategate

Kiwigate appears to match Climategate in three essential characteristics. First, climate scientists declined to submit their data for independent analysis. Second, when backed into a corner the scientists claimed their adjustments had been 'lost'. Third, the raw data itself proves no warming trend.

NZCSC explained their frustrations in trying to get to actual truth about what had happened with New Zealand's climate history, "NIWA did everything they possibly could to help us, except hand over the adjustments. It has turned out that there was actually nothing more they could have done—because *they never had the adjustments. None of the scientific papers that NIWA cited in their impressive-sounding press releases contained the actual adjustments.*"

After a protracted delay, NIWA was forced to admit it has no record of why and when any adjustments were made to the nation's climate data. Independent auditors have shown that older data was fudged to make past temperature appear cooler, while modern data was inexplicably ramped up to portray a warming trend that is not backed up by the actual thermometer numbers.

It is not just in one or two nations that the official government climate numbers are awry. As we are seeing, similar such detailed analysis in North America performed by such esteemed skeptics as veteran meteorologists Joe D'Aleo and Anthony Watts and published in an SPPI paper, *Surface Temperature Records—Policy-driven Deception?*[11], gives cause for concern that we are looking at a worldwide phenomenon.

---

[11]

http://scienceandpublicpolicy.org/originals/policy_driven_deception. html

Antipodean Temperatures Also Faked in Australiagate

It becomes increasingly evident that a case may be proven of a wider conspiracy to commit antipodean climate fraud when we also examine what has occurred in Australia in the controversy dubbed Australiagate.

In February 2010, I published an article, *'Australiagate: NASA Caught in Trick over Aussie Climate Data'*[12] that reported on the findings of two independent climate researchers that analyzed climatic data used by the Intergovernmental Panel on Climate Change (IPCC). The story has been superbly analyzed on two blogs, *Ken's Kingdom*[13] and *Watts Up With That*[14]. The IPCC record showed warming of two degrees per century in Australia that had no scientific explanation. An earlier study by Willis Eschenbach exposing this arbitrary and capricious adjustment was wholly substantiated by citizen scientist, Ken Stewart on his blog, Ken's Kingdom. What was evident was that NASA GISS, based at Columbia University in New York City, had manipulated a century's worth of Queensland's (the Sunshine State) temperature records to reverse a cooling trend in one ground weather station and increase a warming trend in another to skew the overall data set.

But when we look at what the leaked Climategate emails tell us we find that climatologists were conscious of their flaws.

---

[12] http://www.climategate.com/australiagate-now-nasa-caught-in-trick-over-aussie-climate-data

[13] http://kenskingdom.wordpress.com/

[14] www.wattsupwiththat.com

Evidence most pertinent can be read in the 'documents/HARRY_READ_ME.txt' files which can be found at the Climategate[15] website.

These emails address the most recent of the disputed numbers (from 2006-2009) and shows how 'Harry' Harris admits government climate data is unusable: "getting seriously fed up with the state of the Australian data, so many new stations have been introduced, so many false references, so many changes that aren't documented...".

'Harry' then later adds, "I am very sorry to report that the rest of the databases seem to be in nearly as poor a state as Australia was."

These disturbing findings thus call into question both the integrity and the methods of government climatologists and have been condemned by UN IPCC Expert Reviewer, Dr. Vincent Gray[16]. Gray has been a UN IPCC Expert Reviewer for all four UN IPCC reports: 1991, 1995, 2001 and 2007.

Dr. Gray confirms that the raw temperatures, free of the chicanery of governmental 'homogenization' exhibit no such warming bias.

In addition we see that Dr. John Christy[17] of the University of Alabama-Huntsville published two detailed studies that demolished the American 'homogenized' records similarly derived from the same NASA/GISS data sets.

---

[15] http://www.climategate.com/searchable-categorized-climategate-email-database

[16] PhD, Cambridge. Some of his comments can be found at: http://climaterealists.com/index.php?id=6340&utm

[17] PhD, Atmospheric Sciences, University of Illinois. Some of his comments can be found here: http://www.examiner.com/environmental-policy-in-national/global-warming-interview-with-john-christy-models-sensitivity-the-pnas-paper-and-more

Disappearing Temperature Stations

But apart from fiddling the temperatures already in their possession, climate fraudsters sought to manufacture a warming bias in the future by causing the 'disappearance' of 806 inconvenient cooler weather stations around the world. All 806 weather stations were dropped from the total of 6000 worldwide temperature stations in a single year with no explanation from the Global Historical Climatology Network (GHCN), the government organization that maintains this data and which is used by the UN and worldwide governments.

One of these 'missing' cold weather stations is for La Paz in Bolivia and was deemed unnecessary. Now all UN temperature reports come from over 1200 km away. The station at La Paz was at over 10,000 ft above sea level and very cold. Now that the UN ignores Bolivian raw data, the 'homogenized' data for La Paz is suddenly 40 degrees Fahrenheit hotter than before.

World's Two Oldest Temperature Records Disprove Man-made Warming

But when we forego the homogenized government numbers and go instead to the primary source of accurate thermometer readings, we get a different picture with no apparent man-made warming. Two such accurate raw data sets are the world's oldest and most reliable; they are Britain's Central England Temperature Record (CET) and the Central European set from Klementinum at Prague in the Czech Republic. Dr. Jan Zeman, a scientist from Prague, has written a fine paper[18] that proves that there is no human signal in the European record and the overall warming

---

[18] See: http://www.climategate.com/czechgate-part-two-%E2%80%93-the-giss-rape-of-prague

trend in central Europe since the 1790's is a mere quarter of one degree (+0.265°C per century, as shown in Figure 1).

The Prague raw temperatures correlate extremely well with the Central England Temperature Record (CET) that has been running continuously for 351 years (see Figure 2).

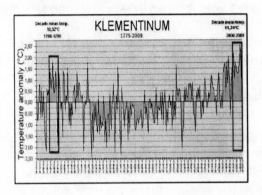

Figure 1: Temperature Anomaly Graph 1775-2009[19]

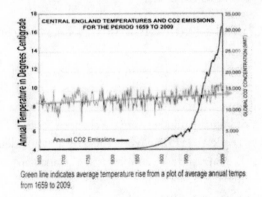

Figure 2: Central England Temperature Record and CO_2 Emissions 1659-2009

[19] Graph by permission of Dr. Jan Zeman.

What we see is neither in central Europe nor in central England has there been any signal of man-made warming in recorded history. As these datasets are considered the best proxies for Northern Hemisphere temperatures and since global temperature trends follow a similar pattern to Northern Hemisphere temps, then the same conclusion on recent warming can potentially be inferred globally.

Rather than publish the facts, the UN's IPCC has instead chosen to misrepresent to international policy makers a monotonous pattern of inexplicable warming by reference to the homogenized numbers—the skewing of which cannot be accounted for from the raw temperature data.

Thus for two decades policy makers were presented with a consistently false picture indicating a warming trend that only existed in the 'laboratories' of climatologists.

Thereby we have identified the true source of 'man-made global warming': it's the clandestine number falsification published in the IPCC summary reports delivered to national governments and world media.

Climate Data Unlawfully Destroyed

Suspicions grew that Anglophone and some European Union governments were faking climate numbers to fulfill a pre-determined goal and their climatologists were being paid to create the illusion of human-induced climate change. Several Freedom of Information requests (FOIA) were filed by independent analysts over several years, most famously by Canadian statistician, Stephen McIntyre[20].

Indeed, Professor Jones of the Climatic Research Unit, University of East Anglia, the world's leading center for climate data homogenization, instructed his colleagues to destroy all such

---

[20] www.climateaudit.org

data and not submit it to McIntyre's lawful FOIA requests. As history now shows, Jones was targeted for criminal investigation due to his unequivocal admissions of misconduct in the leaked Climategate emails. The subsequent official investigations by the UK Information Commissioners' Office (ICO) substantiated the claim that potentially incriminating calculations (metadata) formulated by government researchers in the homogenization process had been destroyed—a wanton criminal act.

Leaked emails written by Jones proved he threatened to destroy his data rather than allow McIntyre to see it...and when the ICO investigated they discovered Jones had, indeed, destroyed the data. Apologists for the crime assert that Jones did not destroy original raw temperature records.

This may be true; however, Jones did destroy his adjustments that would have been key evidence as to his intentions to commit climate fraud. Legal analysts argue the destroyed evidence would likely have proved Jones et al. acted with fraudulent intent. Indeed, statistical forensic experts affirm that if they had been allowed to have examined the data before 'the Jones dog ate it', then any unwarranted adjustments could be readily identified as being caused by faulty system programs or on a one-by-one basis consciously manipulated with the intention to fraudulently deceive.

But, as the ICO found that the evidence had been destroyed during a formal legal inquiry (an FOIA request constitutes a preliminary legal challenge) then the courts are mandated to find that unlawful willful destruction of evidence has occurred. Willful destruction of evidence identified as being relevant to any criminal prosecution renders the destroyer to full liability so that he or she shall be assumed to have destroyed such evidence from a 'consciousness of guilt.' Thereby the court is compelled to render a decision that a cover-up crime has been committed and shall also find the accused guilty of the original offense.

The confession to the crime by Jones is absolute as he provides the prosecutor with both the 'guilty mind' *mens rea* and 'unlawful act' *actus reus*. Moreover, the ICO affirm the reason no prosecution was brought was due to a 'technicality' (an odd statement insofar as there is no time limit to prosecution as per the Fraud Act [2005]).

Unbeknown to those with no legal training, such intentional acts constitute evidence of conspiracy to defraud and Jones should not have escaped prosecution.

My legal associates and I argue the law has not been appropriately applied in this matter. The ICO quite correctly determined that the short statute of limitations (only six months) had already expired making it impossible to prosecute Jones for his crimes under the FOI Act. However, being that Jones confessed to an intention to conspire to break the law and destroy evidence, then on that basis the matter should have been passed to the Serious Fraud Office (SFO) with a mandate to investigate allegations of fraud and conspiracy when sums of money are in excess of £500,000 plus an international dimension is indicated, as per this case. If the British government had applied due diligence then a prosecution under the Fraud Act may still be pursued.

This would have thus been more appropriate to the scale of allegations levied and would also not have fallen afoul of the restrictive statute of limitations that stymied the original FOIA charges.

Private Police Unit Investigated Climategate

For their failures to act, the Home Office and Crown Prosecution Service are complicit in a nonfeasance (failure to act) for not placing the matter into the hands of the SFO, the one department both mandated and particularly skilled to investigate such cases. The investigation was instead assigned to Norfolk Constabulary.

'Aiding' Norfolk Police with their enquiries was privately run secret police unit, the National Domestic Extremism Team (NDET). NDET is directly answerable to the Association of Chief Police Officers (ACPO). Because ACPO is not a public body but rather a private limited company, NDET is exempt from freedom of information laws (FOIA) and other kinds of public accountability, even though they are funded by the Home Office and deploy police officers from regional forces. At the time of this book's publication there has still not been a formal announcement as to the outcome of police investigations first begun in November 2009.

Too Many Coincidences to be Innocent Error?

Notwithstanding the dubious setup of the police investigations, any competent lawyer knows that the accepted legal principle under common law is that when a suspect is proven to have broken the law by covering up, withholding or destroying evidence, then the courts shall correctly apply the 'adverse inference principle' so that he who intentionally destroys the evidence is guilty of the underlying crime.

Since Climategate, the public confidence in government-funded climatology has all but evaporated. Concerned taxpayers are asking how it can be that climate scientists in different countries at the opposite side of the world are facing extraordinarily-similar data fraud allegations. With so many climatologists having 'lost' their calculations, no one can now replicate their methods or have any confidence in the claims that mankind has unnaturally warmed our planet.

The trend is now undeniable; the 'man-made global warming' in the 20th century comes not from the raw data of thermometers readings, but through the 'man-made' tampering perpetrated inside leading meteorological institutions in a handful of English-speaking nations. Judging by the raw temperature data alone and

351

taking into account the shenanigans of a clique of elite researchers, we may safely say there is no persuasive evidence of unusual net planetary surface warming in the last century.

Collectively the authors of this book represent no political ideology or business interest. We are merely enlightened and concerned citizens publishing a own book—arguing a compelling case for a return to reason.

The mid-term U.S. election victories of November 2010 have given the Republican Party a right to claim a mandate to bring an immediate halt to the unnecessary and unpopular imposition of ineffective climate regulations currently being introduced by the 'backdoor' via the Environmental Protection Agency (EPA).

As this book goes to press (November, 2010) *Scientific American* announced the results of their recent opinion poll[21] of its 'scientifically literate' readership. A total of 83% of 5190 respondents think the IPCC (upon which the EPA relies for its science) is 'a corrupt organization, prone to group-think, with a political agenda'.

We offer this volume as evidence both to the U.S. government and other nations so they may act on the incontrovertible facts presented herein and conspicuously discard that mythical Sky Dragon once and for all from all policy considerations.

---

[21] Scientific American Poll Results (November 2010):
http://www.surveymonkey.com/sr.aspx?sm=ONSUsVTBSpkC_2f2c
TnptR6w_2 fehN0orSbxLH1gIA03DqU_3d

Author Biographies

Timothy Ball, PhD (Canada)

Professor Timothy Ball is a renowned environmental consultant and retired Professor of Climatology from the University of Winnipeg. Dr. Ball has served on many local and national committees and as Chair of Provincial boards on water management, environmental issues and sustainable development. Dr. Ball has given over 600 public talks over the last decade on science and the environment.

Dr. Ball came to the fore after his appearance in the sensational British Channel 4 documentary *The Great Global Warming Swindle*. Tim has an extensive science background in climatology, especially the reconstruction of past climates and the impact of climate change on human history and the human condition with additional experience in water resources and areas of sustainable development, pollution prevention, environmental regulations, the impact of government policy on business and economics.

Alan Siddons (United States)

Mr. Siddons is a former radio chemist but now a leading climate researcher and science writer. Mr. Siddons has compiled many chapters on the myriad of errors that have been and still are underwritten by climate alarmists and realists alike. With clear examples and common sense this author illustrates where and why it all went wrong—when the climate alarm bells began ringing.

Joseph A. Olson, PE (United States)

Retired engineer from Texas, Mr. Joe Olson is an internationally-respected science writer with over a hundred major civil engineering and climate-related articles to his name.

Martin Hertzberg, PhD (United States)

Dr. Martin Hertzberg is a long time climate writer, a former U. S. Naval meteorologist with a PhD in Physical Chemistry from Stanford and holder of a Fulbright Professorship. Hertzberg is an internationally-recognized expert on combustion, flames, explosions, and fire research with over a hundred publications in those areas.

Dr. Hertzberg established and supervised the explosion testing laboratory at the U. S. Bureau of Mines facility in Pittsburgh (now NIOSH). Test equipment developed in that laboratory has been widely replicated and incorporated into ASTM standards. Published test results from that laboratory are used for the hazard evaluation of industrial dusts and gases. While with the Federal Government he served as a consultant for several Government Agencies (MSHA, DOE, NAS) and professional groups (such as EPRI).

He is the author of two U.S. patents:
1) Submicron Particulate Detectors
2) Multi-channel Infrared Pyrometers

Claes Johnson, PhD (Sweden)

Dr. Claes Johnson is Professor of Applied Mathematics, School of Computer Science and Communication, Royal Institute of Technology, Stockholm, Sweden and has established himself among the vanguard of international commentators with his highly-respected blog, *Claes Johnson on Mathematics and Science* and *Greenhouse Gases without Greenhouse Effect*.

Charles Anderson, PhD (United States)

Dr. Charles Anderson is a materials physicist with a thirty-eight-year career in the use of radiation to characterize and analyze the properties of materials. He especially enjoys the use of multi-discipline techniques to solve complex materials problems quickly and efficiently.

Dr. Anderson has worked as a laboratory scientist for the Dept. of the Navy, Lockheed Martin Laboratories and since 1995 has owned and operated Anderson Material Evaluation, Inc. He has a BSc in Physics from Brown University and his PhD was earned at Case Western Reserve University—studying the surface magnetization of nickel single crystals using Mössbauer spectroscopy and Auger electron spectroscopy. The analysis of surfaces and near-surface volumes has been a major component of his work since, using such techniques as X-ray photoelectron spectroscopy, infrared spectroscopy, X-ray dispersive spectroscopy, thermal analysis, and light microscopy.

Dr. Anderson has published twenty-nine scientific articles and written thousands of technical reports for his commercial customers. In 2000 he became interested in the claims that man's emissions of carbon dioxide would cause catastrophic damage to the climate and quickly became skeptical. He has written about the many problems with that hypothesis frequently on his blog since.

Hans Schreuder (Holland)

Dutch-born and trained, but now retired analytical chemist and international technical contractor Mr. Schreuder has long been a staunch and highly-regarded critic of the greenhouse gas theory and a leading commentator, using his website (http://ilovemycarbondioxide.com/) as a publishing hub for fellow scientists critical of the theory.

Mr. Schreuder has written many articles on the subject of human-caused global warming and submitted a hundred-page written report, supplemented with a forty-five-minute oral submission, to the Northern Ireland Climate Change Committee in May 2009.

John O'Sullivan (United Kingdom & United States)

Legal analyst and specialist writer on anti-corruption, Mr. John O'Sullivan is the coordinator of these exciting new books and will take the reader on a tour through the maze of confusion surrounding almost all aspects of climate alarm. Mr. O'Sullivan is an accredited academic who taught and lectured for over twenty years at schools and colleges in the east of England as well as successfully litigating for over a decade in the New York State courts and U.S. Federal Second Circuit.

Mr. O'Sullivan is currently Google's most read writer on the greenhouse gas theory (2010) and boasts over a hundred major international science articles. In the U.S. his work is featured in the National Review, America's most popular and influential magazine for Republican/conservative news, commentary and opinion.

Among other internationally-esteemed publications he has appeared in both the *China Daily*, the number one English portal in China, as well as *India Times*, the prime source of business news in India. As a direct consequence of controversial revelations in his 'Satellite-gate' article the U.S. Government swiftly removed a degraded orbital space satellite from service.

ACKNOWLEDGEMENTS

Tim Ball

Thanks to the understanding, encouragement and unwavering support of my family.

Alan Siddons

My deepest thanks go to Hans Schreuder. Being a critic of climatology's current assumptions, I could never have had a better ally.

Joe Olson

I am most grateful to Lady Brewer of Rosehill and Lady Goodbaby of London. Special thanks also to Gabriel Rychert of Climate Realists and Judi McLeod of Canada Free Press.

Martin Hertzberg

For my wife, whose love and understanding were invaluable. Also, to my friends in Summit County, Colorado, and the High Country Unitarian Universalists Fellowship.

Claes Johnson

To Ingrid, for your love and patience.

Author Biographies and Acknowledgements

Charles Anderson

For my beloved wife Anna Palka and my rational, independent-thinking friends.

Hans Schreuder

To the late Dr. Kanner who taught me proper science. To my family and friends who have stood by me throughout. And, I offer thanks and gratitude to Alan Siddons, Gerhard Gerlich, Ralf Tscheuschner, Gerhard Kramm, Claes Johnson and a score of eminent scientists and analysts across the world, without whose insight and encouragement I could not have written my chapters.

John O'Sullivan

For Peggy Clarke and my children, David and Sian. Also, BJ—gone but not forgotten.

Special Thanks

The 'Slayers' team wishes to express particular gratitude to our professional web designer and graphic artist, Thomas Richard, for his stunning visuals and first-rate technical programming. Thomas resides in Boston, Mass. and all business enquiries to him may be made via co2slayer@gmail.com.

蟲椀最甀嚇攀.

LaVergne, TN USA
18 February 2011
217194LV00001B/87/P